FALKIRK C

30124

D1485560

FALKIRK COMMUNITY
TRUST LIBRARIES

CANCELLED

Elephants

Elephants

Birth, Death and Family in the Lives of the Giants

Hannah Mumby

WILLIAM
COLLINS

William Collins
An imprint of HarperCollins*Publishers*
1 London Bridge Street
London SE1 9GF

WilliamCollinsBooks.com

First published in Great Britain by William Collins in 2020

2021 2023 2022 2020
2 4 6 8 10 9 7 5 3 1

Copyright © Hannah Mumby 2020

Images © Hannah Mumby unless otherwise specified

Hannah Mumby asserts the moral right to be identified
as the author of this work in accordance with the
Copyright, Designs and Patents Act 1988

A catalogue record for this book is
available from the British Library

ISBN 978-0-00-833167-2 (hardback)
ISBN 978-0-00-833168-9 (trade paperback)

All rights reserved. No part of this publication may be
reproduced, stored in a retrieval system, or transmitted, in any
form or by any means, electronic, mechanical, photocopying,
recording or otherwise, without the prior
permission of the publishers.

This book is sold subject to the condition that it shall not,
by way of trade or otherwise, be lent, re-sold, hired out or
otherwise circulated without the publisher's prior consent in
any form of binding or cover other than that in which it is
published and without a similar condition including this
condition being imposed on the subsequent purchaser.

Typeset in Adobe Garamond by
Palimpsest Book Production Ltd, Falkirk, Stirlingshire
Printed and bound in Great Britain by CPI Group (UK) Ltd, Croydon

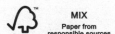

MIX
Paper from
responsible sources

This book is produced from independently certified FSC™ paper
to ensure responsible forest management.

For more information visit: www.harpercollins.co.uk/green

Falkirk Community Trust	
30124 03138796 4	
Askews & Holts	
599.67	£20.00
MK	

Quand je vous parle de moi, je vous parle de vous . . .
Ah! Insensé, qui croit que je ne suis pas toi!

When I speak to you about myself, I speak to you about
yourself. What mad person are you, to believe I am not you!

Victor Hugo

Contents

Prologue

What do you see when you look in the mirror? A sentient being? One with a family and broader social network, a history and memory? An individual with myriad identities, complex relationships fostered over a (potentially, hopefully) long life? Someone capable of expressing emotion, conveying information, communicating with and recognising lots of other individuals and members of other species? Someone aware that they're looking at a reflection of themselves in the mirror?

When I look in the mirror, I see an elephant. That might sound implausible. Let's peer again. Of course I see glasses (or the blur that comes with the lack of glasses) and blonde hair and more spots than a 32-year-old should ever have. I don't see grey, wrinkled skin (at least not to the extent I could seriously be classified as a thick-skinned pachyderm) or tusks. I don't have a trunk, although I heartily wish I did for practical purposes. But we all know that when you look in the mirror, it's not actually *you* that you're looking at, it's a perception of reflection.

The reason I see myself as an elephant is that when you strip back all the packaging, I don't think I'm different to an elephant in many ways. All the questions I posed above could be answered in the affirmative by an elephant if she were asked. You don't even have to ask her; you just have to watch her live her life. In this book, I'm asking you to squint a little bit into the mirror and reflect on the fact that you might have a lot in common with an elephant. I will go through my experiences with elephants, loosely based around key land-marks in the map of an elephant (and human) life. I'll discuss their behaviour, physical changes and interactions with humans they have over the course of their lives. The ultimate aim is to reintroduce you to elephants, not just as majestic and incredible creatures, but also as relatable individuals, friends or even family members, which they become to the people who live alongside them. This is not to lose the science or the wonder, but to give us the tools to rethink our approach to animals and perhaps our priorities in conservation (or just how we define our friendship circle). And to highlight the fact that who we are isn't always as obvious as we might think.

To be very clear, this is not the book I intended to write. Early in the process, as a visiting scholar I sat on an oval of grass at Colorado State University, the bright sun glaring on the screen as I typed on my laptop. It was a devastatingly sunny early autumn day, unfamiliar enough to surprise me and challenge me to confront my barely concealed resentment that a season defined by endings and the spectre of death should be so riotously bright and beautiful. Autumn in Cambridge (the one in England, the one I knew too well) was somehow less contradictory, much more comforting in its gloom. A student approached me and asked what I was doing, and I told him I was attempting to write a 'popular science book' about elephants. He told me it was a bit presumptuous to suppose it would be popular. So, advice heeded, I decided I was trying to write a science book on the topic of elephants. It seemed like an opportunity to convey my ideas to a wider audience than I'd ever had before. I was also excited to go beyond the bounds of my academic writing. After all, there's writing about elephants and then there's writing about actually being an elephant. I know which I think is more fun, and if you can't guess, I'm trumpeting as I write this.

So there I was: Colorado, fall (autumn in my head), my scientific knowledge distilling into something resembling a passable Scotch. Where could I go wrong? Then I wrote something which didn't feel right at all – a chronological and stale list of 'stuff about elephants'. I talked to people and they kept telling me the same thing: I can't see you in this writing and I don't see the science shining through. I had wanted the latter without the former. As with many scientists, I like to take myself out of the equation, because it makes the equation simpler. We are taught to be reductive, to shave everything down with Occam's razor. And in a lot of cases this leads to elegant and rational solutions. But in this case,

I had executed a bit of a Sweeney Todd on myself and haemorrhaged all over the floor. I didn't even make a decent pie from the remains. So it was decided: I would enter the book.

I don't find writing about myself interesting; instead, I often find it embarrassing, narcissistic. I judge myself harshly even as I type. But for the purposes of navigating the jumping timeline of this book, my trajectory becomes relevant. I first became interested in animal lives as an undergraduate, between 2004 and 2007. I have worked with elephants since I was an intern in Kenya in 2010. I continued through my PhD, for which I did fieldwork in Myanmar, and my first post-doctoral position. I was fortunate to receive a series of academic fellowships to return to research in Africa from 2015 onwards, where I built up a team and had the joy of taking my own students on. In 2019, I became an Assistant Professor and moved back to Asia. After a decade of elephants, and a decade and half or so of academia, who I am now is someone wholly consumed with thinking about animals and how people relate to them. It's my job as a scientist, but it's my passion as a person too. In particular, I think about my favourite animals, the elephants, and the animals I have the most complicated relationship with, humans. I think about how elephants interact with each other, with the abiotic and biotic aspects of their environments including us. It's not just thinking, though. I watch elephants and ask people about them. I try to test what the elephants are doing and why, and sometimes I sit in a hide and hope something will happen beyond my legs going numb. I do all of this to the point that I completely lose my sense of whether steamy balls of elephant dung are appropriate for conversation over dinner. I'm going to tell you they're absolutely marvellous for me, but that for other people they might be best reserved for post-dinner drinks, particularly if anything spherical, brown or sticky is on the menu.

I don't think we are very good judges of who we are, which is why I wish I could ask some elephants to introduce me. Unfortunately, I can't adequately put into words all of what they express. For example, on one of my field trips a couple of years ago a young male African elephant stopped to dip his rounded head in the South African dust and then lifted it and shook it about, ears fanned. He was displaying how big he was and perhaps saying I needed to be shown my place. The older males who pointedly ignored me on the same day might have disagreed, ambling as they did past our vehicle in slow motion. To them, I was nothing more than background scenery. Failing elephant explanations, my human family would be the next best when it comes to providing an introduction. My parents would tell you that I was always an odd child. Big eyes, earnest, absorbent as a fancy cushioned loo-roll. I was at home in my head and stubbornly silent much of the time. Occasionally, I had a lot to say. They'd tell you about me asking my dad to film me on a bulky camcorder because I wanted to make a documentary. It was of me on the seashore giving a serious and detailed commentary on the real and imagined creatures I had found there. My parents would point to the flickering footage of that child prodding at the seaweed and they would say it was meant to be – I was always a field biologist.

I shy away from certainty, though. I don't know if I was always meant to find elephants in order to find myself. I don't know if I was meant to be a scientist, a professor, a teacher, a conduit through which a little bit of discovery and knowledge flows, an elephant person. But that is what happened. I think curiosity, luck and fearlessness (sometimes borne of naivety) brought me here. When I think of me, I see myself standing eye to eye with an elephant, then shaking my head, realising elephants aren't great for eye contact and instead holding out my dirty laundry for them to sniff. And I mean my actual dirty laundry. It smells just like me (rather than what I want to smell like), and who better than an elephant, with its magnificent olfactory abilities, to distinguish every layer of the stench. If you are going to give yourself, your absolute on-the-nose flaws and ugliness and all, this is a comprehensive way to do it. And safe too. Elephants are keepers of secrets. A rather big secret being that if you can see past the grey skin and the bulk and the majesty and the fear of them, then sometimes, in some ways, they are just like you. So however I got here, to this place of fascination and science in a joyous positive feedback loop, I wouldn't want to be anywhere else. And I'm going to keep on charging ahead.

Some elephant names have been changed in this text at the request of organisations involved in their care and/or conservation.

CHAPTER 1

Becoming an elephant

There was a time when I thought a great deal about the axolotls. I went to see them in the aquarium at the Jardin des Plantes and stayed for hours watching them, observing their immobility, their faint movements. Now I am an axolotl.

Julio Cortázar

'I can't hear anything.'

'Nothing at all?'

'Honestly, just feedback, wasn't his last point here?'

'Oh, but that point was very early this morning. Let's look for him.'

I sighed and took the radio receiver from my ear. I was tracking a male elephant named Bulumko up in the very north-eastern corner of South Africa with Ronny and Jess, two experienced field workers. I had expected to hear a clear and firm 'bleep, bleep' from the tracking collar which

Bulumko had been wearing for several years now. The collar communicated with a satellite, so we could know Bulumko's movements from afar, and usually with the antenna that I was holding up in the air. But not today. Instead, there was no trace of him, just a tantalisingly close GPS point we'd received several hours ago when he was foraging nearby. I felt mildly exasperated that the technology had let me down. Or perhaps it was just that I hadn't downed enough tea and rusks since we'd set off at daybreak.

Ronny was already scanning the horizon from the canvas roof of our beautiful new vivid green pickup truck, lovingly known to the Elephants Alive employees as Shrek. After tracking elephants from an enclosed *bakkie* 4x4, an off-roading vehicle, we now had the freedom, luxury and coolness of an open vehicle. It was almost autumn, but the daytime temperatures could still soar up to 34 degrees, and Shrek was the best place to be if you couldn't make it to a swimming pool. I clambered up to join Ronny and then jumped back down to get my binoculars, remembering the gulf between my eyesight and his.

Up on the roof, my hunger and frustration slipped away as I felt the precious light breeze fluttering my untucked shirt. I could hear melodious bird calls and buzzing flies attracted to our sweat. After a second, I started to become aware of the landscape. This place could seem vast and monotonous. But after a few visits I had started to pick up on the variation – in parts it was an immense and undulating expanse of flat-leafed mopane trees, joining into a blanket;

in others it was more open savannah punctuated with termite mounds, spiky barked knobthorn trees and the broad-canopied and fragrant marula. I happily noted the subtle gradient in vegetation along the slopes and riversides. It shifted with the seasons too. We'd had a little more rain this year, and there was a layer of green underneath the trees. Not thick, or deep, but somehow speaking of life and its persistence in a place that could be so harsh and arid. I thought of last year, of the blistered, sunburned hippos wallowing ankle-deep in what should have been pools and shook my head to rid myself of the memory.

Ronny's eyes were fixed in the distance. He knew how to track elephants and other animals, so I knew to take my cues from him. He grew up at a nearby safari lodge, where his mum worked in housekeeping, and his uncle was a tracker. Today we were searching for tell-tale signs of elephants: movement in the trees on the horizon or the sound of breaking branches. I could complain about the flashy technology letting us down, but this was really the way to find elephants. You got more of a sense of scale, of how they fitted into the picture; to rely on your senses and be aware of the bush, the sun, and of yourself within it all. Having said that, I'd made plenty of phantom elephant sightings, mistaking rocks, tree trunks or buffalo for one of our big grey study animals. Today, we were going to get the real thing.

Suddenly, Ronny smiled. 'There he is! In that big block of trees!'

I held up the binoculars to my myopic eyes, a result of

too many hours looking at screens and books close-up and not enough out here. Of course, Ronny was right. Bulumko. He was a towering and impressive bull in his prime. He was old enough to have distinctive notches in his ears. He probably caught them on branches or thorns, and elephants tend to have more ragged ears the older they get. We used them to confirm his identity. He'd also grown a pair of jutting tusks, by no means the biggest I had seen, but each over a metre long. He had been tracked for years by Elephants Alive; and knowing the elephants as individuals with their own lives and experiences over time was what attracted me to them in the first place. What's more, the elephants had names, not just numbers. Bulumko's name meant wisdom, but what always struck me was the way he moved. He ambled, feeling the trees with his trunk and taking time to determine his route. He favoured travelling along the dirt roads that we drove along in our vehicle, and he occasionally lingered by water sources longer than the other males. Like every elephant I had ever known, Bulumko had his quirks and unique features. But he was a little different even beyond that. Bulumko was blind.

In many species, a blind individual just wouldn't survive. But sight isn't as important for elephants as it is for other animals. Hearing and smell are much more critical. Bulumko has not just found a way to navigate the world without his sight, he'd also found a way to navigate social life. He could displace other males for access to the freshest water or a prime shady spot to rest in. Even if he wasn't receiving direct care

from other animals, he was certainly accepted and even domi-
nant because of his stature, age and tusks. Our relationship
with Bulumko was singular, because while the other elephants
we watched became habituated to our presence, Bulumko
accepted us as friends and co-travellers. We could stay with
him for hours by a waterhole as he bathed and we enviously
sweated in the sun. When he moved off, he rumbled, a short
but low and sonorous sound indicating to us to follow – let's
go! I often wondered whether Bulumko was more likely than
other elephants to communicate with us in this way precisely
because of his blindness. With him, we were beyond being
observers, we were almost another elephant. Perhaps not
seeing us made it easier for him to interact with us as companions, and our car was like another big animal. Except it smelt
of people and talked like us too.

Excited to 'talk' to Bulumko again and, of course, to listen,
I clambered down off the roof, eager to drive to a spot where
we could get a better sighting of him. I smiled at Ronny
and Jess. 'This is the life!' Jess laughed and shook her head
at me. But it really was what I wanted, to be treated like an
elephant by an elephant. Then Jess revved Shrek's heavy
engine and we drove to Bulumko, leaving a trail of dust in
our wake.

What is an elephant? A favourite character in a childhood
book, a weapon of war, a religious icon, a draught animal, a
pest, a conservation flagship, a source of income, a drain on
income, a hunting prize, a tourist trap, a terrifying beast, a

gentle giant, a source of ivory or a source of power. It has at one time or another been all of these, and the diversity of responses illustrates the complexity and history of our relationship with our world's biggest land mammals. Few animals evoke such heartfelt passion and division of opinion. It often strikes me that these enormous animals carry with them so much meaning, symbolism and sometimes baggage, but at least they have the strength for it. Many scientists have to carefully introduce and describe their study species of cichlid fishes, fruit flies, honeyguides or sticklebacks. When I say that I work with elephants, most people already have a firm image in their minds, which can be hard for me to compete with. But the way I see elephants is as personal as it is for everyone else. For me, elephants are the biggest, most complex and endlessly fascinating puzzle I have ever faced. And I've done one of those thousand-piece jigsaws. How elephants can make evolutionary sense with their long lives, inefficient guts and slow reproduction in a world where being small, reproducing fast and multiplying exponentially is an option continues to intrigue me. How do we have giants in this world? And how do we live with them?

It wasn't inevitable that my life would end up this way, defined by elephants: their lives, their movements, their habits and the similarities and differences between us and them. I didn't grow up in the bush like Ronny. I'm not South African like him and Jess. If you'd asked me when I was seven what I wanted to be when I grew up, I'd have said Indiana Jones or David Attenborough. But these early field-based dreams drifted away as I was increasingly buried in academic literature, revision and exams. I feel as though I came to from my teenage years as an undergraduate student at King's College, Cambridge: slightly baffled as to how as a first generation university student I managed to end up there, but equally determined that I had been very sure it was where I should be. I was overwhelmed by the looming architecture, at once oppressively cloistered and staggeringly grand. It echoed my inability to engage with both the nuanced and visually spectacular traditions, acted out by others with an ease and naturalness I couldn't hope to emulate. I was equally intimidated by the wider features of the student experience. I didn't row in one of the college boats, I didn't act or write for the newspaper, I didn't get a glitzy internship. I sat in the library, shell-shocked by how very little I knew and could ever hope to know.

Feeling adrift, I needed a framework to understand myself, as an individual and as a human being. I decided the solution to understanding my place in the world was to get to the bottom of what it means to be human. Unsurprisingly, even this task was a little over-ambitious for my undergraduate dissertation. Then, in a second-year lecture, my Cambridge experience turned around dramatically. I found the ideas of life history, and, with it, an elegant clarity that comes with the most beautifully constructed scientific theories. At its core, the theory is about explaining how living organisms balance energy and time in their development, reproduction and senescence.

For me, it was about understanding any living organism by

getting to the nuts and bolts of how it is born or comes to being, grows, reproduces (or not) and dies. These are universal traits that bind us living things together, but we experience them as individuals. This means we can observe enormous variation both between organisms and between lives; for example, the overall picture is that humans are long-lived animals with slow-paced lives, but this masks the fact that some of us live a few minutes or hours and others live for over a century. How do these patterns of mortality put pressure on the shape of our lives as a species? It was the perfect lens through which to see myself and the world. It allowed me to put aside the idea of humans as exceptional and use the principles to understand my own life as just another creature.

With this in mind, I started studying the lives of primates. Non-human primates like monkeys, apes and lemurs are the go-to comparative organisms for humans. This makes complete sense on one level, because they are our closest relatives. I launched myself into conducting an ambitious study for my undergraduate dissertation. I was trying to use a life-history formula to determine whether the pace of human life was typical of a primate, or if it was an outlier, standing out from the usual pattern. What I mean by the pace of life isn't just the lifespan. I mean whether key 'milestones' in our life history, such as weaning and age at first reproduction, are very spread out, or if they are clumped together. It's often useful to view life histories as on a continuum. For example, in mammals it could go from a 'fast-living' creature like a mouse, which matures rapidly, has many offspring and a relatively short life, to animals that have a longer development, space out their offspring more and have a longer life, like humans. All of this can be influenced by body size: large bodies take longer to grow, so we have to take this into account. Without celebrating myself too much, I was excited to discover that although humans are living in the slow lane even among primates, we humans aren't an absolute outlier. We aren't even the slowest of the species I researched.

I felt thrilled with my contribution to scientific knowledge, and the head of the Anthropology Department recommended that I publish it. I felt accepted and valued in a way that I hadn't in several years of port and cheese parties (although I had managed to develop a taste for port and a fear it would give me gout, and a particular joy in truffled pecorino). Then, I thought more about it, and more, and then just too much. I realised how limited my study really was. I had a horrible, sinking realisation that I was effectively standing with my nose a couple of inches from a pointillist painting and happily telling everyone I could see the big picture. I could in fact just see some lovely green dots, but I could definitely not see a picnic scene by the River Seine (or whatever your favourite pointillist scene depicts).

At the time, I was working on an alumni database at a Cambridge college. Twiddling my thumbs, romanticising about becoming an academic, but not ready to inhabit that life. I began to unpick my failure and re-evaluate my approach. Really, I had put humans into their primate context, but humans aren't just primates. And how much do we have in common with more distantly related primates like lemurs? What I needed to do was forget about relatedness and just look for the slow life-history pattern across animals. If I could find an animal that had evolved on a separate trajectory and wasn't related to humans, it would be all the more interesting. I could investigate the similarities and differences between lives without having to worry that I was just describing the general primate pattern. I started a list. I wrote down 'whales'. I grimaced at the thought of seasickness and crossed it out. I wrote down 'elephants', and changed the direction of my life.

A few months later I arrived in Kenya. I had never been on a flight with request stops before, never mind on such a tiny plane. A stroke of luck meant that the former master of the college where I had been working had been an influential elephant researcher in the 1960s and 1970s and had recommended I seek out a PhD student he had examined. The PhD student was Iain Douglas-Hamilton, who had become a giant of elephant research. Before I had even met Iain, I had been told he was an 'elephant'. I knew by reputation that he was superlatively intelligent, incredibly productive and passionately committed to his work. But I thought that being an elephant sounded mad. I had heard about him being chased around a thorny bush by a female elephant and being lucky that he had only ended up with a few scratches. The real mystery, he told me, was her intention. 'Did she actually intend to kill me, when she overran me and plunged her tusks nine inches into the ground above my head? Did she change her mind after I was under her, at her mercy, or was it only her intention from the start to frighten me? Either way, she was certainly successful in increasing my heartbeat!'

Despite these fortuitous connections, Kenya was more alien than Cambridge had ever been. When we landed in Samburu I was in equal parts happy I was alive and proud I had kept down my breakfast, particularly because one of the younger passengers had not managed the latter. I hauled my backpack off the plane and dragged it through the sand to the edge of the runway. I glanced at the rickety table of knick-knacks bearing a 'duty-free' sign, unsure as to whether it was a spirited attempt at a shop or a joke. The airstrip was deserted. I sat on my backpack and wished I had water. The sun was getting higher in the sky and I squinted. I thought I should have reminded someone I was coming here. Could it be they didn't know, and I was going to have to get back on that terrifying little plane? Luckily, my anxious musings were broken by two men asking where I was staying.

'Save The Elephants,' I told them.

The representatives arrived half an hour later: two smiling young men, Jerenimo and Benjamin. I felt uncomfortable that I must have been several years older than them and asked myself if I was being ridiculous by following them, but I realised my alternatives were non-existent. They didn't seem the least bit uncomfortable. On the contrary, they were grinning at me. We chatted in the vehicle as we bumped along the sandy roads. My overwhelming first impression of Samburu was of toasty orange sand and thorns. Thorns so impressive that I didn't want to think of stepping on one. I retreated into my safe space, scientific knowledge, and reflected that plants must have to strongly defend themselves against plant-eating animals to evolve those five-centimetre protective prongs. While I was lost in my thoughts, Jerenimo stopped the car. He was still smiling.

'We have to walk the rest of the way.'

I looked at the river that he had parked beside and the remains of the bridge spanning it. Benjamin told me that the river had flooded earlier in the year and destroyed the bridge and much of the camp. It had plenty of crocodiles, so we were better off climbing over what was left of the old bridge. Gallantly, Jerenimo and Benjamin took my bags. I wished I had a better sense of balance, another thing I should have been focusing on instead of all that reading. I began coaching myself. *Hannah, this is fieldwork. You said you wanted to do it. You want to understand elephants? You need to live in a tent, go on those dirt roads and climb over this bridge!* I convinced myself that I was not an attractive meal to a crocodile, without giving the idea the scientific scrutiny it might have merited at other times. I climbed clumsily up the side of the bridge and peeked over the top.

Baboons!

The flashing canine teeth of the larger male baboons impressed me much more in the flesh than I had expected. I knew they were mainly for display, but I ducked back down.

Jerenimo and Benjamin put their knowledge of animal behaviour into action and pulled themselves up to their full height, puffing out their chests, clapping their hands and shouting out loud. The baboons duly dispersed and I made a mental note to behave more like a baboon when I was around baboons.

A few days later, I had settled into camp life. Despite having been devastated by the flood that had destroyed the 'Baboon Bridge' just five months earlier, the camp was tidy, well-equipped and functional. The days quickly took on a form and rhythm that became familiar, and I realised that I could more easily adapt to living anywhere than I had imagined. I was staying in a tent underneath a corrugated iron roof, to protect me from the vervet monkey excrement dropping from the tree when I woke in the morning. I had an orange plastic bucket to wash in, which a small antelope with a spring in its step known as a klipspringer had also taken to using as a fresh water supply. I would go to the water pump every morning and fill up a transparent bottle. I would leave this in the centre of the camp, exposed to the sun, in order to warm the water for my afternoon bucket shower. I became used to the long drop (even thankful for the breeze it created) and to making sure I put the rope across the entrance so that people knew I was in there. I played checkers using the tops from beer bottles; the gold Tusker brand tops got me thinking of the symbolic power elephants have over us. As well as the curious klipspringer, I met some of the other animals that frequented the camp: a wizened pair of hornbills, still striking in their senescence with white and black feathers and curved orangey-yellow beaks; a band of mongooses hunting by the breakfast table. The troop of baboons I had first met on the bridge appeared calmer as they traversed the slope next to our breakfast table undisturbed. There was a group of rock hyraxes, fluffy mammals that look a bit like oversized guinea pigs but are actually some of the closest living relatives of elephants. They slept a lot, and sometimes took a break from basking in the sun to investigate the bags of flour in the food store. But however charming these characters were, I hadn't yet seen an elephant. After all, that's why I had come all the way to Kenya.

The following morning at breakfast, I got much more than chai, toast and honey: my first wild elephant sighting. To this day, I can't imagine a better gateway to watching elephants. I can't begin to explain how lucky I was that it happened when I was on foot and in the camp. The former because it allowed me to sense the scale of the elephant in front of me and my own fragility, which is just not possible in a vehicle. And the latter because I was familiar enough with the camp not to let the experience terrify and overwhelm me. An old male elephant, Yeager, had come to investigate the old collars that had been worn by elephants to mount the tracking devices. The delicacy with which he handled these bulky and decaying pieces of abandoned equipment touched me. He explored them with the tip of his trunk and turned them gently, detecting the scent of their former wearers, faded over time. I remember his face so clearly. It was wide above the trunk, as happens to males with age, with deep horizontal wrinkles traversing the front of his face. Either side, his eyes were looking downwards, the direction of his gaze emphasised by the dramatically long and spindly eyelashes shading them. His tusks were thick at the base, another sign of his age. His right tusk splayed out to the side and formed a stubby but impressive point, while the left one had a messy jagged break about 5 centimetres down from his lip line, exposing the layers of hard tissue and giving him a rakish, rugged appearance.

Yeager moved so slowly, but with so much power and intention. I watched him pull up grass with the tip of his trunk, flick the trunk back and forth to remove the dust from the roots and place the grass in his mouth, chewing methodically and rhythmically. I don't know how long I watched him for. It could have been minutes or hours, but it didn't matter because it was on his time. The pace of my perception slowed to match his, and with it some of the urgency of this world diminished: the irresistible urge to push forward, even when I didn't know why I was doing it. As he eventually eased his way out of the camp, he paused to defecate, dropping five balls of caramel-coloured dung. When he was safely out of the way, I walked over to where he had been, taking time to look at his footprints in the sand, with their criss-crossing cracks like trails on a map. From the flat area at the back of his foot I could work out the direction in which he had been walking. I could see where he had rested his trunk on the ground, leaving a valley of ripples. I smelled his dung, grassy, warm and not the slightest bit repulsive. I squeezed it between my fingers, letting the green-yellow water run down my arm. There was life in that dung. I felt the barely digested plant material, gathered from kilometres away and deposited here; the potential for dung beetles to roll it away, or a frog to take up residence there, or the seeds in it to germinate. I thought of the worm eggs that I'd only later see under the microscope, and, smaller still, his DNA, the bacteria from his gut, the metabolised fragments of all his hormones. I saw him at once as more than the sum of his parts, a vital part of a system that was both much bigger and much smaller than him.

I had come to Kenya looking for a slow-living animal, a model on which to test my ideas. I laughed at myself for dismissing Iain as an elephant before I had met one myself. What Yeager gave me was the experience of feeling my own life slow down, when I watched him and his face being seared onto my soul. I knew I wanted to spend the rest of my life working with elephants. But for me, it wasn't enough to watch in awe and appreciate. My love and science had collided again, just as they had done at the first life-history lecture I attended. Yeager raised more questions in me than he answered. The biggest scientific issue for me was that perhaps my theoretical framework had been too reductive: by considering individuals as the embodiment of their life-history landmarks, I had lost too much of their complexity. Although Yeager was alone, I saw so much interaction – the grass he ate, the collars he sniffed, that incredible dung he dropped. For me, this crystallised the fact that elephants in isolation from their environment make no sense. I knew elephants had startling parallels with humans in terms of their life history, but broadening what I could imagine, see, test and understand about them might reveal even more intriguing parallels and contrasts. I couldn't study them the way I had originally intended, with elephants just as a model. By doing that I would limit them, limit Yeager because he was so much more. I had to understand what it was to be an elephant and try, in the limited and human way I could, to conduct my studies from that point of view. A big evolutionary, ecological puzzle like an elephant deserves more than just our fascination, it deserves our critical thinking. That way I could enhance both my experience of being human and my understanding of being human on every level, from the deeply personal to the scientifically challenging.

I am a human. I am an elephant. The study of life has separated humans from other animals with no particular scientific justification for this exceptionalism, other than that as humans we might be both most familiar with and most intrigued by our own species. Many beautiful and reasoned arguments have been made against human exceptionalism (I refer you to anything by primatologist Frans de Waal), and there are wonderful accounts of individual animals exhibiting behaviours that we might once have thought exclusive to humans, eliciting responses and understanding concepts, from honey bees to scrub jays. What doesn't seem to have changed much is the way we do science. When we study humans, we are ethnographers: we place ourselves within the context that we are interested in and participate with informants trying to understand what they are doing and why. When we study other animals, we try to make ourselves invisible and remove ourselves from the context, just as I had planned and monumentally failed to do in the writing of this book. Why? Is it because we think we are a distraction, that we will alter the behaviour of the animals? Perhaps, in part. But there is a huge wealth of material on reflexivity, positionality and the role of the researcher in social science that natural scientists could tap into. Acknowledging the presence of the researcher, and all of the complication that comes with that, actually makes the studies richer. I think part of the problem is our deeply ingrained sense of difference from other animals and it is holding us back as scientists. If we were more free, if we were willing to rumble right back to Bulumko, what world of understanding could it open to us?

CHAPTER 2

Samburu families

I come as one but I stand as 83
That includes my ancestors and my living family.
Dope Saint Jude

Up early one morning to do a mammal census in Samburu, we turned the corner and followed the bend of the Ewaso Ng'iro River. Although the sun was low it was already warm, a sure sign it would be baking later, and I sat out in the open on the back of the vehicle. I was half tempted to lie back and bask in the sun, stretching my arms along the rails and tilting my head towards that vast African sky. But thoughts of the thorny bushes we negotiated our way around as we drove along – on top of that irritating prickly heat one gets when mixing pale skin, doxycycline for malaria prophylaxis and the intensity of the sun – stopped me. Besides, I should have taken a cue from the mongooses and been a little more alert: the mammal census was important work. Essentially, we drove a transect line, not a straight one but zigzagging along the dirt road that ran through the landscape of Samburu and the neighbouring Buffalo Springs Reserve, across the river. On our trip, we recorded every large mammal, bigger than a house cat or so, we saw along the way. This was a handy task for me, as it allowed me to start to recognise the different animals and begin to understand how they lived. And as an added bonus, it allowed me to shout out 'dik-dik' (the name of a small antelope), when we saw that abundant species at regular intervals, so that Jerenimo could record it on our sighting sheet.

I'm fortunate that I learned to tell an impala from a Grant's gazelle on the mammal census drives. I also learned that the Grevy's zebra has ears like a mouse, unlike the horse-like ears of the common zebra. I learned what a savannah was. How had I managed to live for so long without caring about those details? Even more lovely, on that morning, I had another special sighting, of the royal family bathing in the river. Victoria, Elizabeth – I saw them all. Of course, it was a group of elephants named after that famous family who I saw having a bit of a bath and a frolic. The royal family are one of the largest and most frequently sighted herds of elephants in the area. Each herd in Samburu, consisting of a matriarch, her female relatives and their offspring, is named according to a theme. It was just by chance that I got to see the royals having a bath, but it gave me an extra chuckle. Having had the meeting with the big male Yeager in the camp the day before, I felt lucky that these were only the first of many exciting sightings to come in Samburu.

The elephants were doing a bit of their own thing while being very much together, as elephants often do. I suppose this is group life, family life, and something that I do with my own family all the time. Being a female elephant is about that togetherness and family; this is the framework around which everything else in her life is built. Elephants, especially the females, just don't make sense in isolation. As I watched the royals, some were standing aloof, knee-deep in the sediment-tinged water. The river was still swollen enough to make recent droughts seem like a distant memory. From time to time, the elephants dipped their trunks into the water, sucking it up to splash across their bodies and letting it fall in rivulets down their wrinkled skin or blasting it straight into their mouths. Others waded into deeper water, almost completely submerged with trunks raised aloft, until just the tip of their trunks broke the surface, like prehensile snorkels. I noted the contrast between the different parts of their body; the areas that had been submerged or splashed had glistening wet skin, so different from the normal dusty dry skin. But all of it was crisscrossed with deep lines, and all of it pachyderm thick. I reflected that the shiny wet skin spoke of a different kind of mammal – they almost seemed like manatees, one of their closest living relatives. In some ways, elephants seemed to fit better, visually and behaviourally, in the water than on land. The water gave them an elegance they never seemed to achieve in their terrestrial lives.

ELEPHANTS

I smiled as I remembered Aristotle's writing about elephants – he got a lot right, such as the plant-based diet, and the importance of the trunk. But he got their swimming ability wrong. He said that they couldn't, and anyone who has seen elephants in water would question this. Now we know they don't have to have their feet on the riverbed at all – they really do swim. I also noted my typical academic attachment to reading the work of such authors. Had I really thought Aristotle was the best preparation for coming to this place? Self-flagellation aside, watching the babies was a delight. The youngest were still small enough to duck underneath their mothers, a good indication that they were less than a year old. Their trunks were still fairly novel to them, whipping around like errant noodles and without the care and precision one notes in older elephants. Other young elephants were bigger and bolder, splaying out their ears and chasing egrets in the thick mud with their trunks outstretched. After the egrets flapped and moved away, the calves returned to their mothers, swinging their trunks.

Through sightings like these, I collected data as well as gawping, grinning, and trying to snap that perfect photograph of the moment before a spray of water hits an elephant, spreading for a second like a glistening diamond necklace that shatteringly disintegrates when it hits muddy skin. I learned to distinguish the males from the females, noting the sharper foreheads of the females when viewed from the side; the level of their heads, matching with their shoulders and backs; and the swollen or looser breasts hanging between the front legs of the females that had calves (which was almost all of them) or were expecting them. When an elephant was partially obscured, perhaps by a bush, or the elephant had its back to you, there were other things to note. In one female, I saw that distinctive skin that hangs straight down between the back legs, around the openings to the urinary and reproductive tracts. Males had a more rounded head, higher on their shoulders, and a face that widened with age. Like Yeager, the base of their tusks became thicker, from appearing like sharp white thorns in youth, to branches and then to fully blown tree trunks in the oldest males. Sometimes they showed their penis, which is spectacularly large, but visually reasonably unremarkable, just like that of a horse. However, if you watched for a while, you'd notice it is prehensile, they can move it around with a lot of control, and seeing a male elephant scratch his belly with the tip his penis for the first time was quite an experience. But more commonly, I would just see the curved cover that starts between their back legs, reaching towards their tummy, the penis tucked inside. And they had no visible balls to compete with the ground squirrel in the Save the Elephants camp who sat on his pair while he nibbled his food. Elephant testicles are internal.

Nevertheless, it became quite easy to spot the sex differences, and I noted that I rarely saw a male taller than the shoulder height of a female in a herd. You see, this royal family had no Charles or Prince Philip or Henry VIII. Males have a different relationship with their families because all the males disperse, either gradually or by making a clear

29

break. Some are pushed a little and some venture out on their own. Either way, when they are teenagers they leave the herd they are born into; destined to move on, only hanging out with the family groups to follow a female in season. In contrast, the females remain with their mothers and female relatives for their whole lives, bound by those familial ties.

The female family relationships reminded me of my own family. I was sure we would look much less grand having a bath together, but my grandmother, mother, sisters, nieces and I could compete with the elephants for noisy greetings, a consistent stream of close-range vocalisations, intermittent touching, cooperative herding of babies and frequent stopping for snacks. As a social animal, I felt a kinship with them across the species divide because humans are family animals too. But with that empathy, of course I noticed the differences. Whereas elephants mature with males and females living separate adult lives, humans are more like many of our primate relatives living in mixed groups. But there's not only one way of living as a human and we vary hugely in how much adults interact with each other, and with their families. Elephants have the pattern of male dispersal across their ranges.

As a scientist watching the elephants, I was interested in their whole lives – the shape, duration and events over a lifespan. And with these animals, this was something that felt very female-focused and female-led. I saw a female life mapped out before me, the various stages being performed by the family, from the calves just a few months old to the matriarch – a life of energy demands, punctuated by phases of growth, sexual maturity, successive reproductive events, ageing, death. The social life seemed to slot nicely around that: mother-offspring relationship, bonds with age-mates, the persistence of social bonds, becoming a leader with age.

The temporal stability of the female social bonds, so tightly intertwined with familial relationships, was incredible to observe. Elephants can live for decades – into the fifties or sixties is not uncommon, and up to eighty is not unheard of, particularly in Asian elephants. As in most mammals, life expectancy is higher in females than in males. Matriarchs, the de facto leaders of elephant families, capture a lifetime of social, spatial and environmental knowledge. Almost a decade after this encounter in Samburu, I was asked to talk about female leadership in elephants, and I considered giving a two-sentence lecture: female elephants are leaders because the structure of their society means there are no adults males around. They're not female leaders, they're just leaders. Then I could let everyone go for lunch early. Back in a real elephant herd, the males leave either of their own accord or with some pushing from an older female, perhaps to avoid inbreeding. Adult males don't often come back. I know to look for them when a female is in oestrus, following at the back. Or sometimes I note when a dispersed teenage male returns, taller at the shoulder than his mother and aunts, to walk with his family group for a while. I genuinely don't know if it would be useful to apply that elephant society structure to how humans interact. And while I still ponder this, years after meeting the royal family, the elephants just carry on living their lives.

The matriarchs are leaders because of their age and experience; there's very little questioning among the other females. For example, I watched the royal family on that first mammal census day as they were bathing: when it was time to move off, 'let's go' was rumbled, a short but sonorous vocalisation that has elements below our hearing frequency, but you can hear the harmonics. It was emitted first by the matriarch and then echoed through the adults in the group. And the elephants moved off, all in the same direction, with the matriarch stationed purposefully at the front, striding a little and causing some of the younger ones to pick up their pace. Within a couple of minutes, they had all left. Those family bonds tied the group together and kept them moving in sync across the landscape. No one was left far behind. What had moments before been like a family swimming pool session, the kind you see in humans with a shallow pool for kids, a wave machine and maybe one of those spiralling waterslides, was now just a muddy patch of river, the banks pocked with footprints and the odd bum impression. It was quieter now, perhaps more crocodile friendly. We in the field vehicle weren't anthropologists anymore, we were more like archaeologists – looking through our binoculars at a space that had been inhabited, with just what was left behind to piece together what had been. Despite being huge, elephants seem just to disappear, and sometimes the matriarchs are the magicians.

SAMBURU FAMILIES

Around the time I was just getting to know elephants in Samburu, my colleague Lucy Bates was already much more familiar with elephants and doing some fascinating studies, further south in Kenya, with the famous Amboseli population, where over 2,500 elephants have been studied for decades. Samburu, with its orange-tinged baked earth and studded by hardy acacia, will always hold a special place in my heart and in elephant research. But Amboseli is also supremely important. If you have ever seen a photograph of a large herd of elephants walking tall in grassland, with the snow-capped silhouette of Kilimanjaro in the background, it was probably taken in Amboseli. It's one of those images that makes you utter 'majestic' under your breath. And why not? What's incredible to me is that generations of researchers in Amboseli have been conducting meticulous and extensive research in this beautiful place for over forty years. They must spend less time gawping than I do.

As I watched herds in Samburu, I thought of one of the studies Lucy had undertaken. Lucy and her team presented female elephants with soil mixed with urine from other female elephants. If I did this to you, with human urine, you might find it unpleasant. However, you have probably noticed how important scent from urine is to other species. I myself have spent more than a few minutes standing, as patiently as possible, while my dog sniffs a urine sample from another dog, or from some other animal. After filling her nostrils, she often deposits her own sample too. What Lucy was interested in was whether the elephants would react differently to familiar elephants, their kin, and those that were unfa-miliar (and unrelated). She found that indeed they did; they showed less interest in the urine of unrelated individuals who they're not used to being around. This might seem unsurprising, but Lucy took it one step further. She also tested reactions to urine of kin that were walking ahead of the elephant in their group, so the elephants might expect to come across their urine in their path, and contrasted them to samples of urine from kin walking behind them in the group. The first female passing a sample from an elephant walking behind her reached out her trunk and sniffed towards the sample of urine more than she did towards that of a non-kin elephant or a family member walking ahead of her. She would also do this if the sample came from a family member who wasn't currently with the group. The reactions Lucy saw were subtle but fascinating. An elephant could not only distinguish between individuals based on urine alone, but also exhibited knowledge of where family members were located in relation to herself. It felt very elephant, being so aware of the rest of the group.

I suppose it's something that we do all the time, just using a different modality. If we thought our family member was just behind us and then we heard their voice ahead of us, or caught a glimpse of them, it would be confusing, and we might take some time to investigate what was going on. Interestingly, in her experiment, Lucy compared the way that elephants made trunk movements towards the samples to the longer looking times which human children exhibit before they can speak, when their expectations are violated. In their understanding of other individuals and the relationship to them, of family and the familiar and unfamiliar, it's hard not to compare the elephant responses to what a human might do – indeed, so hard that I was tempted to write 'our understanding' in the previous sentence. I don't necessarily think that this comparison is a bad thing, as long as it doesn't become an exercise about trying to force an elephant into a human mould. It wouldn't fit, not simply out of sheer physical size, but also because an elephant isn't more or less than a human. It's just different enough that, in order to see that elephant in my reflection, I have to squint a bit.

In contrast to females and families, my interest in the lives of male elephants grew much more slowly, fostered by a few leading researchers. For example, Kate Evans worked in another incredible ecosystem, the Okavango Delta in Botswana. I'm not sure if it's a coincidence that most elephant research sites are in such beautiful places. Perhaps there is a correlation between that beauty and the remote locations in which elephants are still able to live in this anthropocentric world we've created. Or perhaps elephant researchers have a bit of an eye for a radiant high noon, a lazily glowing sunset, a sinuous river, that one baobab tree that stands out against the sky. Any words I have for the Okavango Delta could hardly do it justice. I'd been lucky enough to visit, almost exactly six years after meeting those elephant royals in Samburu. I found myself in another part of Africa, arriving on another dusty runway. But the Delta is not my parched Samburu.

The Delta is water. It's inland, fed by the Angolan highlands, seasonally fluctuating in water cover and far more beautiful than these little facts can convey. Flying above it, gripping the back of the pilot's chair with one white-knuckled hand and digging my nails into the palm of a fellow passenger with the other, I glimpsed the Delta from above and cursed myself for being such a wretched flier. The landscape – let's call it a waterscape because that's why it buzzes with life and otherworldly beauty – is immense. If two million hectares means anything to you, that's the kind of size we're talking about. But when you're in that little plane, it just means that you can't see the end of it; in such a dry country, you see water where a mirage should be, weaving streams between islands, papyrus and reed rafts drifting on the flow. Both the most frightening and the most exciting thing about those little planes is that you can see the detail below: you count the animals from far above, you can see the elephant in the river, that herd of antelope kicking up water, the island in the shape of a heart.

I stayed on that heart. I thought it suited me. A little bit of symbolic sentimentality imposed on something far more complex, both messier and more magnificent. Since my days bumping around on the back seat in Samburu, I had been lucky to have many drives in the field, but the Delta trip was my first boat trip to observe animals. Something about being on the water made the shared space feel reduced, and the experience was somehow less zoo-like, more engaged and intimate, but also calmer than being on foot. The very best was travelling by *mokoro*; being paddled along in a canoe-like boat removed the engine noise, the dust, the sense of being part of a pretty impressive machine, and left me feeling smaller, slower, more aware of every frog clinging to a reed and insect skating across the water surface. Everything felt more elegant. The pace, the smoothness, the low-lying boats all made sense. We were closer to the eye level of a wading hippo, closer to being less of an observer, more of a participant. As we drifted past a baboon sitting atop a termite mound, his tail dark against the sky, surveying the waterscape as the sun sank below the horizon, perhaps I was the one being observed rather than the other way around.

While Kate was working in this place, she did some research on male elephant adolescence that made me rethink how interesting male elephant life could be. Males, just like females, grow up surrounded by their mother, perhaps their grandmother and lots of female relatives. They suckle, play, grow up. But they had fallen off my radar when they dispersed from the family herds and then returned, fully formed, as big males like Yeager. Kate's study filled in some of the gaps. She found that adolescent males (in the age classes 10–15 and 16–20 years) were the most sociable age groups, with preferences for larger groups and being close to other elephants. They didn't like to be alone. So female elephants and young elephants with their family herds were not the only ones that liked company. In a way, adolescence for male elephants was more of a major transition than for females, because not only were they just becoming sexually mature, they were also navigating being away from their family, forming relationships in a different context and relating to the environment differently too. Kate also found that the older males (aged 36 years and over) were the preferred social partners of all age classes of males. She suggested that, as in females, older males might act as repositories of social knowledge. In the Delta, I came across a big old male named Jack that I couldn't help thinking might have made for an interesting social partner. The first signs of him were healthy dollops of dung deposited on and around the paths to our tent in camp. Thankfully, the owners of the camp realised that the elephants had used the same pathway to access palm trees long before they set up tents for humans and still allowed them access to the route. Like with so many routes in the Delta – the hippo highways between floating islands, the termites building the mounds that cemented some of those floating islands into the riverbed, the fish eagles fighting in the sky – we let the animals take the lead and shape the space.

It was palm nut season. I hadn't realised that there could be such a wonderful thing. Our tent was at the end of the elephant path and closest to a cluster of tall makalani palm trees – the kind that spring from all the permanent islands of the Delta, skinny-stemmed and fan-crowned. Within twenty minutes of our arrival, Jack was at the trees. Through the mesh at the side of my tent, I watched him lift up his head and trunk and push both of them firmly against one of those skinny trees. He shook a whole tree, firmly and forcibly, seemingly impervious to the loud swishing rustles of the leaves as the tree swayed and the nuts rained down, some bouncing off his head and body as well as falling straight to the ground. But he didn't remain uninterested in those nuts. After about forty seconds of solid shaking, he stepped back and inspected his bounty with the tip of his trunk, scooping up the nuts into his mouth and chewing rhythmically. After seeing Jack's method, I noticed some elephants using his 'trunk up' method and others using their foreheads to push and shake the trees. Others lacked Jack's impressive stature and were not able to generate the force to rock the whole tree, so they were compelled to find different solutions, picking up fallen nuts, or even swiping a few from a larger elephant who was still busy shaking a tree. This variation hinted at learnt behaviour, perhaps socially from observation or from trial and error. If one elephant observed the technique of another and replicated it, that was special, because it represented the potential for culture. It was one of those traits we thought of as distinctly human until we realised that being human had blinded us to its existence in crows and chimpanzees and macaques. So we had to redefine being human and place ourselves on a different pedestal, a shakier one, with wobbly legs just to maintain a special place in our own collective imagination. I felt impressed (with Jack and his species) and exasperated (with myself and my own species) at the same time.

One day, as Jack was lying about fifty metres away, I took a good look at him – a sleeping giant, his chest rising and falling so slowly, in that slowed down elephant time, and one of his tusks jutting out, a deep groove towards the tip marking its use in breaking branches. On the pathway, I prodded a pile of his dung, a mound of barely processed palm nuts surrounded by a little grass, greenish-caramel in colour and looking very healthy. He had basically just transported the seeds and coated them in fertiliser. Perhaps they would grow into trees, one day to be shaken by his descendants. I thought again about Kate's research, and how my view of big male elephants had shifted. They weren't just shadowy figures, disappearing from families as adolescents and living a solitary life. Jack had his own social network. He might have been an important source of knowledge for younger males, but his life was just a little bit harder to trace, and his family might have been about more than kinship – and perhaps that was a little bit more of a challenge for me as a scientist.

Back in Samburu in 2010, family and challenges were at the forefront of my mind. In a short amount of time, my colleagues and the animals in camp felt like a family. On my second day, I woke up to a rucksack being thrown into my tent. Anyone who thinks of themselves as the baby, the one who needs most looking after, is familiar with the feeling of an unknown newcomer threatening to take over that status. But I was very lucky. The newcomer was Heather Gurd, a fastidious researcher with a passion for conservation, an excellent knowledge of wildlife and a tolerance for my lack thereof. She was generous, easy company and allowed me to maintain my status as 'most clueless'. She also lent me lens cleaner for my glasses. There was no sibling rivalry; we quickly became friends. I got used to taking it in turns with Heather to have a hair wash in a bucket (sometimes after several days of hat wearing to cover our greasy locks) and to wash my knickers in a different one, hanging them on a line like bunting over my tent. To me, the time I spent chatting with my hair wrapped in a towel and my drying undies hanging overhead played just as much of a role in my feeling at home as the game drives. I was just in a new home, with a new normal family.

My routine was hardly opaque, and the vervet monkeys seemed to know all of my movements. They formed a large group of males, females with babies clutching their backs and plenty of youngsters who were acrobatic, lively, and swung from their tails in the trees. I can't say I enjoyed their raucous company. My stepping out of the tent was a great time for them to defecate from an overhead branch; my absence on a game drive would be a good opportunity for them to try to open the tent zipper and find out if I still had some sweeties stowed inside. I felt I was some clumsy and less agile primate cousin that they were thoroughly trolling. But the monkeys had their uses after I got wise to the zip tampering and wrapped sticky tape around it to make it more difficult. I could listen out for the monkey alarm calls – there are specific ones for an aerial threat like an eagle, and different calls for a snake or a leopard. The camp was open to all of those creatures, with no fences, and the monkeys were not just aware of what I was up to, but also of all the other inhabitants and guests. They were a filter with lots of eyes and lots of voices, and although their constant activity could be distracting, it didn't take much thought to tap into it and experience the benefits. My human colleagues were also better at spotting snakes than I was, whether the snakes were resting in the kitchen or under a desk in the office.

I never did manage to avoid being shat on by the monkeys when I emerged bleary-eyed from my tent in the morning. I suppose that's what families are like: they'll save you from that snake, but they'll also shit on you from time to time, or at least have a good laugh when you've been crapped on.

An elephant family might not be too concerned about the monkeys. However, they certainly interact with faeces, primarily in the form of calves eating the dung of their mothers, as in many other herbivores. What scientists delicately term coprophagy is just eating poo. Okay, that seems a lot even for close family, but that dung contains bacteria that could aid digestion; and when a calf eats the dung of its mother, it's giving itself access to all of those handy bacteria, lining its gut and investing in a healthier digestive future. Isn't that one we'd all like? Perhaps we're the ones late to the party with this, because recent research on faecal transplants in humans has shown that this can really help patients who have experienced severe gastrointestinal symptoms. And when it comes to group responses to risk, like the monkeys and their alarm calls, the elephant families have their own methods to keep the herd together and safe. Elephants do have predators – lions can hunt baby elephants, some humans present a threat and, famously, although not predators, elephants aren't fans of bees. They react to bees as a group, clustering with the babies in the centre, with older females facing outwards, trunks raised: a defensive and impressive formation, and an interaction dependent on communication and inter-species knowledge. Even though they have thick skin, they can be stung on sensitive areas such as the tip of their trunk. And the younger elephants might be more vulnerable because of their softer skin.

Everything in Samburu was brand new for me, but some things were just brand new to the world. The idea of family support and communication was at the forefront of my mind when one day we came across a young mother and her very young calf, a little distance from the rest of their herd. The calf was pinkish all over with translucent ears like lettuce leaves, indicating it was only one day or so old. It was small, slipping easily below the mother and having to lift up its front legs to reach her nipple and suckle. Its skin was baggy, giving it the look of a child who has been given clothes to grow into. It was cute, no doubt about it, but there was also something uncertain and a little fragile about the scene. The mother was a little tense, wary of us and even of her calf. It was probably her first calf, given that she wasn't particularly tall herself and had none of the hallmarks of older age. Her face was full, not concave, and there was no sign of the long saddleback you see in females as they get older. Although I could have watched the little elephant all day, the scientific cogs in my mind began whirring into motion. This family life and group life being so tightly linked made perfect sense in terms of evolution. The related females share a lot of genetic material, as do their calves, even though they might be fathered by unrelated males. Oftentimes, the matriarch is the grandmother of little calves like this. So it would make sense for the mother and her offspring to be supported by the rest of the group, because it's improving the chances of the matriarch's genetic material to be carried into the next generation. Helping the female would be a boost to the outcomes for the calf. And of course, it was beneficial for her mother too. This experience would provide her with a direct descendant and give her the vital experience for her next calf. I smiled as I took a parting picture of the calf, its eyes open looking in our direction and mouth wide in one of those expressions you'd like to interpret as a grin. Without more experienced females and support of a family, the future would be very precarious for this calf and its mother. I felt comforted to know that there was this safety net, that it made perfect evolutionary sense and that I could sit back and watch it.

But camp life wasn't all idyllic multi-species interactions and elephant observations. Snow White singing to bluebirds we were not. I don't mean to suggest I didn't appreciate it. I liked everything from the superb starlings with their brilliant blue necks and orange breasts to the shudder of excitement that comes with sightings of crocodiles in the river. I always looked out for crocodiles, those intimidating 'Jurassic logs' you know you don't want to disturb, but which are somehow rapturously exhilarating when they do rouse from their ligneous state. Mainly when their craggy mouths are stained with blood, and they catch your eye for a protracted moment in which you hear only your heart thumping. I can fall into that eye, right into the vertical pupil, for as long as the gaze holds, until there's a blink, three eyelids sliding shut with reptilian grace, and the crocodile slips away again, to become live driftwood. No, there was a darkness for us in Samburu that didn't come from crocodiles, but from a predator much closer to home. It came from us.

It all started through the crackling of the radio. We were used to news about elephants, perhaps the location of team members, or even hearing about a fantastic sighting – a leopard lounging in the tree with an easy elegant leg draped over a branch, a lion nonchalantly licking between its toes in the noon sun and showing you the most impressive feline footpads you've ever seen. But this news made a tension settle across my family. An elephant had been reported dead, presumed shot, presumed illegally, just a few kilometres beyond the bounds of the reserve – our safe haven, our home, our place where animals were a part of life. At breakfast, we picked at our bread and honey. Chris, a very experienced tracker and researcher, drove Heather and me all the way through the park and out of the gate on the western side. We weren't laughing today about what dik-dik sounds like. We were solemn.

On leaving the park, we crossed from the bumpy Samburu sand trails to the glossy straight tarmac road. The road was quiet, with more people walking along it than driving. We drove through the nearest town, Isiolo, and beyond, out into the bush again. Chris knew where we were headed, and Heather and I knew we had to go, even though we didn't want to; it was somehow important to witness it. The thorny bushes took on a more sinister appearance as we crashed through, looking for a carcass. There were no vultures to guide us, but we didn't really need to look – it was the smell and sound that knocked me off my feet. The corpse was buzzing with flies and hazy with a rotting stench. I gagged several times and tried to swallow, burning the back of my throat with acid and disgust. What was lying on the ground wasn't an elephant anymore. It was a grey faceless mass squirming with maggots. It didn't seem as though it could have been an elephant a week ago, although that's what Chris estimated. He took samples to identify the elephant because a lot of the features – trunk, tusks, eyes – were gone, hacked away. We might need tissue and blood to identify this elephant using DNA. It was an adult female, Chris told us. But it didn't look that way. It looked like a melted candle of an elephant. The ears slumped, and there was an open wound where the face should have been. The feet had been removed, and one lay forlornly a couple of metres away, those beautiful elephant toes disconnected. I was too weak for this. My stomach certainly was, and I felt the vomit rise again, this time from really deep. An image of it steeped in bile came

46

into my head, and I imagined heaving black blood like the clotted patches on the elephant remains. I stepped back, for some cleaner air and clearer thoughts.

But it only got worse. Chris pointed out the hyena tracks and even a dead hyena that had probably been killed in the scramble for access to the elephant. This was a crucible of death, but instead of being vivid, it felt dull; the colour was drained out. Everything was drained. I swallowed my vomit again, and the death stench wrapped around me. Within months of seeing this fetid corpse, there would be many terrible reminders of the importance of family to elephants. Orphaned elephants with no older female relatives were left behind, and in other cases, the team would find the body of a baby elephant by the mother. While the first elephant corpse I ever saw was garishly gruesome, these calves could be heart-breakingly intact and peaceful – a swollen infant that could merely be sleeping if it weren't too stiff, too still, too close to the buzzing flies reproducing in the remains of its mother. The bond between mother and baby was simply too strong, so the calf stayed close and died there: another victim, less bloody, even more tragic.

I don't remember the drive back to our camp, but I knew something had shifted. We were all feeling some kind of loss. Mine was of innocence. Experiencing dead elephants changed me as much as knowing living elephants. It's not all frolicking in the water, chasing egrets and following the matriarch, or Yeager coming to camp and showing us what stature really means. It is blood in the sand and the shockwaves of an elephant killed by humans, rippling from the maggot-ridden epicentre to the swollen dead calf. It is the rest of the elephant family, changed forever, somewhere out there in the bush. It is the herders that found the body. It is the people who had fired the gun and hacked off the tusks. It is the tusks, now called ivory, in transit to a port by now. It is the researchers and the wildlife service working fastidiously under the toughest conditions. The world has lost an elephant and acquired one more giant grey scar. I went to bed without having eaten much, fearful of nausea rising again.

This was 2010, and I wish I could tell you it was an isolated event. But I had caught the uptick of a spate of elephant killing that bloomed and spread across the African continent, tracking the vulnerability of elephants, political instability, corruption and desperate people, a spate that some experts would argue is still continuing now. The killings were illegal, the type that we often refer to as poaching but which is much more than that. It is the sinister and bloody tip of pernicious iceberg-like networks of crime and exploitation. Of the tens of thousands of elephants killed, each instance was horrendous, the sheer numbers masking the pain of each individual death. There are not enough words in the world to respond to it adequately, to articulate the pain of the whole system, certainly not enough scientists with a tendency to throw up to express the stomach-churning horror of it. My nausea didn't feel like enough. I felt hollow.

Since meeting my first elephants, I felt like I was living with a new duality of elephant and human. I thought about whether I could really be elephant, or had the millennia of evolution separated us so much perhaps that I could only get a flicker of a connection? Humans killed that elephant, and mourned it, sold its parts for money, and protested that sale, bribed and were bribed, arrested and were arrested, campaigned and fought against killing elephants, killed people who killed elephants, killed people who protected elephants. Humans bought tusks, and both impressed and disgusted other humans with their purchase. Humans ignored the whole thing or noticed for a second and then got on with their lives. Perhaps duality was out of the window when there were seemingly so many humans contradicting each other. I couldn't grasp how I fitted into it. Even with my grief, my youth, my naivety, I knew it would be too simple to frame the humans in this story as good and evil. We're multi-faceted, adaptable, capable of feats of great beauty and generosity and horrific acts of violence and destruction. Oftentimes, we don't operate at those extremes at all or form small parts of something that only reveals itself under a much wider lens. Elephants are complicated too. Elephants are not exclusively victims; they are strong and powerful. They can be protective, they can attack, they can kill each other, they can kill people. Of course, they haven't organised themselves in the same way that humans have. But our individuality, our complexity, our seeming contradictions that prevent us from ever being captured in a trope like 'gentle giant' make us similar in some ways. Perhaps the layers of being human can let us see the layers of being an elephant and not limit them to fit what we want them to be.

Even after the death, I didn't want to leave my Samburu families, either the elephant groups or the camp, but I knew it was part of growing up as a researcher. I felt like an elephant, and sometimes elephants have to go out on their own, the way males leave their natal group. And I wanted to know more about elephant lives now that I knew about how elephants make sense in their families. I had this idea that if I could quantify the lives more thoroughly, I could find some kind of demographic formula to stop elephants teetering on the edge of extinction and make a better world for them. It was reductive and simplistic of course, but I wanted to make a difference. And I was twenty-four, so forgive my over-inflated sense of my own role in the world.

I got the whole team together on my final morning and we posed on the vehicle, hanging around and off it like monkeys. Our big smiles would be interpreted as aggression in the vervet monkeys but were meant to show happiness in humans, although we sometimes smile with our teeth but not our eyes if we are sad. I climbed over the remains of the flood-ravaged bridge one last time. Jerenimo held my hand as I clambered along inelegantly, and Heather snapped some parting photos. I was so grateful, so changed, so addicted to elephants, so driven. I held two contrasting images in my mind: the fresh newborn with pinkish ears, but also the dead baby elephant. Both were almost smiling: the first with skin too big for it, the second with internal bleeding and oedema that made it look bonny and plump. I didn't look back.

CHAPTER 3

Oozies and elephants

A story always sounds clear enough at a distance, but the nearer you get to the scene of events the vaguer it becomes.

George Orwell

Around two years later, I found myself just outside of Katha, a town in northern Myanmar, where George Orwell was stationed as a member of the colonial police and where he set his novel *Burmese Days*, which was his great foray into purple prose. I could see why Katha inspired Orwell. It was a bustling market town tucked along the ribbons of the wide Irrawaddy river. In front of me, a neat row of eight young male elephants stood to attention, facing me straight on. But it wasn't intimidating. They stood still, then lifted up their trunks and showed their pink tongues. I smiled. My tongue has a bit more freedom of movement in my mouth and I thought about sticking it out at them before deciding against it. The elephants were on the move now, still in formation. The wooden bells tied around their necks clattered joyfully as they turned to walk in a line. Then they all delicately traversed across a series of tree stumps, the kind you get in one of those adventure playgrounds for kids. These very big, but still immature, animals navigated them with much more grace than one would expect for creatures already on their way to weighing a tonne. Some of the elephants were smaller than that, only five years old and about 500 kilogrammes, their shoulders not higher than my head. I clapped as the last elephant stepped down from the stumps. They were not putting on this show alone, though; either a teenage boy or young man wearing a bright blue tracksuit and baseball cap sat atop the head of each elephant, or more or less where the head meets the neck. These were oozies, elephant drivers and caretakers. Each was directing the elephant using spoken commands, pressure on the back of the ears with his feet and sometimes his hands on the elephant's distinctive double domed forehead. But it didn't look as though much pressure was required. The boys could stand up on the elephants, balancing easily, or fall asleep on them, pulling a baseball cap over their eyes. From my viewpoint on the ground, the relationship appeared easy, well matched. A young human and a young elephant, performing, being together, reflections of each other. Although the show was for me, this was all about them. After the showing off, the boys leaned gently and nonchalantly next to their elephants, holding them by an ear as they kicked at the ground with their toes and chatted. The grass was thick and green in the opening of the forest where we were standing, and beyond the clearing, tall trees bursting with foliage surrounded us. The place, the humans and the elephants – everything was different here.

I had ended up in Myanmar because of an advert in the journal *Nature* for a PhD studentship. And this wasn't just one of those generic ones inviting students to compete for funding; it was specific. It was to work on Asian elephant life history. I was quite sure I wouldn't be competitive. Think about how many people see an advert in *Nature*; I bet all of them like elephants. But after my experience in Africa with elephants like Yeager, I was determined not to let this opportunity pass me by without at least trying. I applied and somehow, I was invited to attend an interview. The morning of the interview, I was taking antibiotics for my acne and took one pill too many. I felt queasy. I vomited copiously in the train loo on the way. Why is my elephant experience, my life experience, always punctuated this way? I remembered I had also vomited before my Cambridge undergraduate interview and that my mother had smeared a bit of lipstick on my cheeks to give a bit of life to the ghost. Seven years later, I was pale again and my stomach tied and re-tied itself in knots. I took a series of trains, trailing slowly from Cambridge to Sheffield University, the host institution of the studentship. I looked out of the window at the unspectacular fields and gardens. I wasn't sure about Sheffield, but I pulled up to it eventually, proud in its post-industrial, semi-redeveloped 2010 glory. I ate some anaemic sushi and stared at the naked February trees.

Then I had one of my blind luck moments. I can't help grinning as I type this. I was about to meet Virpi Lummaa, who was then a research fellow and was in charge of the project. I don't remember much of any consequence from the interview, other than that I told her, 'I used to be a little mousey girl, but now I'm not.' She threw back her head and laughed, as if I had said I used to be a teapot. Afterwards, I phoned my father from the train outside Alfreton to tell him it was hopeless and also to ask where Alfreton was. I had hardly made any progress at all on all counts. The journey home dragged. I got back to Cambridge and lamented the floppy sushi and stomach-acid-tainted performance. After a suitable moping period, I checked my emails, and everything turned around. There was one from Virpi. She wanted to work with me. Start when you can. There's an official procedure and so on, but I want you to study timber elephants, the elephants that drag felled trees to the rivers in Myanmar.

What Virpi gave me access to was data, the shining building blocks of scientific discovery. In this case, the data related to whole lives of generations of elephants from Myanmar, a country I had never visited, but had looked at it on a map and saw it was shaped like a kite flying between China and India, the string trailing down along Thailand into the Andaman Sea. My knowledge of Myanmar was incredibly limited. I knew George Orwell had lived there because I was an Orwell nut. I knew of Mandalay and Moulmein as romantic places from poems and songs, seemingly unreal, swathed in the kind of mythical grandeur and lyricism that evoke El Dorado or Camelot. It was overwhelming and confusing, but I was going to this place to understand the lives of elephants. I would use something extraordinary: little green books compiled by generations of Myanmar veterinarians that contained the records of births and deaths, as well as details of the life and health of elephants working in the timber industry.

In order to find the elephants, I first went to Yangon, the old capital and thriving cultural centre. The Shwedagon temple dominates the centre of this city, with its colossal stupa covered in gold and surrounded by elaborately decorated shrines, walkways and smaller temples. The city was somehow bejewelled and incredibly glamorous despite the potholes, sprawling suburbs and lack of street lighting. Women wore flowers in their impeccably styled and braided hair, and they wore perfectly fitted silk or cotton longyis and blouses with gold thread and sequins and black velvet slippers. I don't know how they managed to stay looking so stylish as they glided between street vendors, navigated cracked paving slabs, and stepped pristinely from buses that had never been near air-conditioning. With my sheen of sweat and perpetually slightly swollen ankles, I felt I was rather letting the side down. But I gamely tied my hair with a bow and had some bright longyis made, with their black-banded waistlines fitted to accommodate my post-food poisoning waist. Longyis are long pieces of fabric similar to a sarong, but made as a loop you step into. They are worn by both men and women, but tied strictly along gender lines with a central knot for men and a fold to one side for women. They did not allow for my usual wide-pacing gait. Some of my longyis sported regional patterns, such as the narrow horizontal stripes of Shan or the distinctive multi-coloured diamonds of Kachin or vivid red, black, green and yellow of Chin. Needless to say, I can't fit into them at home, although just looking at them brings me joy. Luckily, I had only so much time for the distraction of glitzy dressing, since I had my treasure trove of data to analyse.

Three young Burmese men carried the huge ledgers into the breezy flat in Yangon where I was staying. Virpi had recruited several veterinarians and helpers for the project, including the maternal Khin Than Win, who had taken me under her wing and provided a lot of in-country expertise. The large ledgers were composites of the little green books for each elephant. They listed the names, ID numbers and dates of birth and death of elephants going back to the 1920s. I tried to be as respectful to my Myanmar colleagues as possible and kept topping up their glasses of refreshingly cool lychee juice. One of those men later told me he had thought I was a teenager helping to look after the apartment and I wasn't sure whether to be pleased or not. We opened the heavy books, each one almost a metre wide, and leafed through them on the wooden floor. It felt a little unreal. I found myself nodding as the veterinarians pointed out the high number of elephant deaths in the period of World War II and that the records covered the periods before and after the country's independence in 1947. Here were the lives of so many deceased elephants, yet many were still alive and owned by the government: around 2,700 as I flicked through the records in 2012. It was a story of Myanmar told through elephants and I felt how integral that human–elephant relationship was to this country, even though I didn't yet have a grasp of it at all.

My first trips to Myanmar in 2012 are more vivid than any of the others I took over the next three years. There was something about being thrown into it that made me remember all the details, and the later trips were all a well-intentioned but ultimately unsuccessful attempt to recapture the magic of the first. I didn't even know how to say thank you (it's 'kyay zu tin bar dae'), but I had a lot to be grateful for. The first time I drove with Virpi, Khin and a motley crew of scientists, veterinarians, and people with cameras to record the event, we went first to Mandalay, on the smooth, wide highway from Yangon. I wanted to tell everyone I was on the road to Mandalay, but it was a small minivan and people were quite tired of it by the time we passed the 50-mile marker. It was April and the ground was dusty, the temperatures searing hot, but I felt pure joy the whole way. I don't know what it was; perhaps the idea of the elephants somewhere up there. Perhaps it was something else, less tangible. Or it could have been the discovery of lahpet, a pickled tealeaf salad with peanuts, beans and garlic, which is much tastier than it sounds. I do have to advise you to enjoy it in Myanmar, though. I once travelled with it and it arrived in London wilted, shrivelled and inedible, and vaguely resembling some herbal product it wouldn't be advisable to cross borders with.

We had to move quickly, eager to get to our appointment with those elephants. And this was when I felt that I really saw Myanmar. The roads were sandy, mottled with holes and narrow, which I preferred to that endless highway-like-a-wide-runway between Yangon and Mandalay, because of the lives played out alongside them. There were stalls where vendors sold coconut water, others where leaves of bright red and addictive betel were rolled and wrapped, strung together like unholy rosaries for people to buy ten at a time. I tried it once. I chewed slowly, looking uncomfortably sanguine as my saliva turned bright red. I couldn't contain it in my mouth and hung over an open drain, spluttering and spitting, flicking bright red liquid on my clothes and toes. It felt inelegant. And then the heart-pounding buzz like five double espressos kicked in, and I felt restless, fidgety, and counted the pulses of blood rushing through my ears. I am not made for this, I repeated to myself, stop pushing yourself so hard!

I loved looking at the shops along the roads. In towns, the buildings might have more permanence. But once we were beyond the Mandalay suburbs and the sister city of Sagaing across the Irrawaddy, we saw stilted houses, of bamboo wood and palm leaves. I looked into teashops selling breakfast food such as parathas with chickpeas and tea sweetened with condensed milk and stewed to suit the palate of each customer. There were adverts on these establishments, for soap, for beer, and for coffee that comes all-in-one with milk and sugar and you just need to add hot water. And just as in Yangon, the people were impeccably dressed, embarrassing me in my khakis and sweat (actual sweat, not a sweater) combo. Well-groomed children waved to us as we bumped past, their faces decorated lovingly in fragrant light yellow thanaka, a kind of make-up and natural sunscreen made from the bark of a tree ground to a paste. Many women and children I saw wore it, and I took to doing so myself, as I found the sensation of it cooling in the hot months. I liked the way it made me feel when I was taking time and care over myself even when we were staying in some pretty rustic accommodation. What's more, I liked the smell – like sandalwood with a bit of citrus.

When travelling early, we saw long lines of monks carrying their bowls in the soft morning light, a perfect canvas upon which to project our ideas of Myanmar. Later, children were walking to school in starched white and green uniforms, square-shaped embroidered cloth bags in bright colours with two strips of fringe at the bottom slung across their bodies. But the most common mode of transport we saw used was definitely the motorcycle, although we saw minibuses crammed with maroon-robed monks and all manner of improvised agricultural vehicles. Whole families piled on motorbikes or scooters, and there were people transporting their baskets of fruit, vegetables or just more baskets, their chickens, goats or fridges. What a place to drive through at 20 miles an hour! It wasn't all wonderful though, especially not when I was at the very back of the minibus with its brown patterned seats straight out of the 1970s. The combined stiffness and shaking abilities of the vehicle gave it a real immediacy and made me feel like my organs would no longer be discernible from each other, instead forming something approaching the consistency of a blended iced cocktail: too dull in colour for a strawberry daiquiri, but in bad lighting, you might mistake it. (There were no daiquiris to be had in Myanmar – the only real booze around was something that was called whiskey but smelled more like paint stripper than scotch, and the traditional palm toddy, which had the scent of coconut water and a memory-erasing potency that might knock you off your feet and leave you unable to ride side-saddle on a motorbike in a longyi. Khin was most concerned about me consuming it, and of course, I have no recollection of ever having done so.)

We finally ended our crawl northwest of Mandalay and arrived in the Alaungdaw Kathapa Park late. The elephant place. It was dark, and the driver carefully negotiated the route as we wound uphill. Then he suddenly jammed on the brakes. We peeked out and saw the reflection of the car lights in the eyes of two adult elephants standing in the road. My first Myanmar elephants. In the dark. I didn't expect this, these faint outlines of nocturnal beasts, much more challenging to see than in the bright Kenyan sun on encountering those first African elephants. The driver didn't expect it either; he balked and hit reverse and we zipped back, about 200 metres down the road. He talked at length to the driver of the vehicle behind us, eventually being reassured that these were captive elephants and should let him by. I still look at the blurry photos I took of that first encounter, of those elephant shadows with flash eyes. Close, but completely unclear.

The next morning, I had a more idyllic encounter. I walked down to the river and watched as an older oozie washed his elephant. The light was so soft it almost looked misty, but the air was dry. A circle of trees grew tall, well over 15 metres next to the shallow river, branching out into spindles and providing the most beautiful frame for my photographs. Yes, Hannah, I thought, this is it. This is when you get to think about the human and elephant being one in this place; that drive-by last night was just a bump in the road. The oozie wore a checked shirt, backwards baseball cap and a sheathed knife attached to his belt. He was balancing on the back of his elephant, cleaning its ears. The elephant turned on his side and proffered a foot, which the oozie duly cared for too. The oozie bunched and tied up his longyi to make a kind of pair of shorts. He scrubbed the elephant diligently, including his one tusk. It really was a different way of humans being with elephants, less of the majesty perhaps than I had seen in Africa, but vital for me to see. It reminded me that elephants in service means humans working like this with them. Myanmar is much less densely populated than other parts of south and southeast Asia, but perhaps one of the reasons elephants have been able to persist in the region is because of this close relationship with humans. Not that all elephants live with humans, but where they are so vital for an economy – for their many different values, spiritual, martial and more – maybe there's more space for them.

I felt the weight of history in those glimpses of the country as well as in the elephant records. The green book data on elephant lives had been collected for over a century, ever since the country was a British colony and the economic value of each elephant was huge because of the might of the timber industry. Before scarcity of hardwoods and without mechanisation, elephant power was the only way to get big beautiful teak or other hardwood logs from the forest to the river. From there they would float all the way downstream to towns like Mandalay, where they'd be processed. In the heat of March to the beginning of the monsoon in June, when the river level dropped, the forest crisped up and gradually baked to a golden brown, and the elephants and oozies took a rest. These powerful machine alternatives could nimbly tackle slopes and walk between trees, so there was no need to clear fell and build roads. In return, they were able to forage freely in the forests, in a similar habitat to wild elephants, and the government employed specially trained elephant veterinarians, each responsible for around 100 to 250 elephants.

The workloads of every elephant were recorded, and their working lives strictly regulated. They could work for several hours, five days a week, but would then be released into the forest to feed and socialise. Some of the elephants would be hobbled on chains. The females would get a year of rest after giving birth to a calf, during which time it would always be with her, free to suckle, then it would follow her to work until it was trained and paired with an oozie at the age of five. At the age of seventeen sexually mature, but not yet of adult stature it would become a fully-fledged working elephant until retiring at the age of 50, along with its oozie. Around 10 per cent of males and 30 per cent of females reached this retirement age, during which the government was still responsible for their health-care and access to food and water. This is far more than reach the same age in western zoos. It was a highly structured regime with the aim of maximising productivity and efficiency in terms of timber extraction, but it also created a kind of welfare state for elephants. Each of those working elephants in the whole country had, and to this day still have, a name, a number and a little green book. These green books are filled in by specially trained elephant veterinarians and carefully stored by head oozies, the leaders of the teams of elephant handlers and riders, until the end of the elephant's life, when they are filed away as part of a huge database.

Because of this meticulous and long-term record keeping, we can track the lives of elephants for up to five generations; and for those born into captivity, we know their mothers, grandmothers, great grandmothers and so on. For an elephant to whom families, female kin and the accumulation of experience over long lives is so important, these records were ancestry, social history, epidemiology and real-time census data all in one. But perhaps I didn't see all of that potential at the time, as my scientific study of life history took a very specific perspective. I was interested in how evolution shapes the life course and key events in the lives of organisms so they can optimise their evolutionary fitness, which we often measure as the number of descendants. But in time, records actually allowed me, and the whole of Virpi's team, to conceptualise elephant life history in a much broader way, including the narrative records of birth stories with details of the height of each calf born, bold elephant escapes and recaptures, and how these elephant lives related to the lives of other elephants around them. All this was documented in notes indicating mating sightings, where each elephant lived and worked, and any medical problems it encountered. Long after I took flight and left Virpi's nest, her group continued to make fascinating discoveries because of this dataset.

On a different level, the records even helped me to make sense of this country, with its complex history a backdrop to the elephant lives and all of the humans who worked with them and filled in the books. Frankly, it's easy to see an oozie as just the boy or man (I don't think I ever saw a woman who was an oozie in Myanmar, although there have been some). A man who sits on the elephant and directs it, teaches it commands and feeds it a banana as a treat. But they are much more than that. They're the people with the elephants every day, who interact with them most, and because of the social and family focus that elephants and humans have, it's not surprising that they form very close bonds. The oozies bathe their elephants in the river, use distinctive bark as soap to wash them, massage them with herbal compresses, and ensure their safety when dragging those logs. The elephant doesn't work without the human. When we wanted to collect blood samples on a field trip, samples that could be used to look at hormonal, inflammatory and genetic markers, teams of veterinarians and oozies rubbed the side of a nervous elephant and sang a soothing song to her. I wanted to join in. I felt relaxed. Later, one of the vets told me it was a song they sang to elephants when they were being trained. Separated from their mothers and learning to form a close bond with humans, for those elephants it was part of growing into maturity. And the memory of being rubbed on the side and sang to meant this technique could always be used to calm them, even decades later.

It wasn't always idyllic, with one species looking at the other and understanding everything. That is an injustice to the individuals who can definitely have their challenges. Some oozies didn't enjoy working in very remote sites, some realised they could make more money working in mines, some didn't have the best relationships with their elephants. It's not always easy communicating across the species divide. But there was something very different from the experience with African elephants. There's a real closeness with captive elephants in this kind of free contact with humans even when it's scary; because sometimes it is scary when an elephant is agitated or sick or distressed, or maybe one of the humans is. There's something about being in it together that can make it feel very visceral. One elephant, upon observing the blood sample collection we were carrying out on another elephant, reacted quite strongly. Perhaps he didn't much like the look of it. He took off, bounding along with his oozie gamely dangling next to him and clinging on to his ear. Just as being in Samburu taught me about seeing elephants, Myanmar made me see how close humans and elephants could be.

Alaungdaw Kathapa, the park we were staying in, isn't really known for timber elephants. In fact, most of the elephants working there were not working in logging at all, but worked for the Forestry Department, transporting tourists uphill to the shrine to the disciple of the Buddha after which the park is named. I decided to follow the pilgrims and ride with Virpi and the rest of the crew. However you feel about riding elephants, it was a fascinating experience. (I'm of the view that it's something that has been done for a long time in some places and that there is a huge spectrum of ways in which it can be done, in terms of safety, comfort, risk and health, for elephants and for humans.) We climbed up onto platforms before getting onto the elephants and rocked our way up. I was reminded of how agile and elegant elephants can be. I weigh a fraction of what they do, and yet I am not as sure-footed, nor as able to delicately negotiate a slope. From our position, I felt the size difference. The world looks different from an elephant perspective; a lot of it to be looked down on. But the oozie was comfortable on his perch, and after we got off, he fed the elephant some cooked rice. Shin Maha Kathapa, after whom the park was named, was a hermit; he had been the third follower of the Buddha, and a close disciple. After the death of the Buddha, Maha Kathapa took up the solitary life and was later buried in a cave. And it was at this cave that the elephant dropped us, alongside a large temple. I remember going down into the cave, clutching a couple of tiny squares of gold leaf to put on its wall as is so commonly done by Buddhists at sacred sites like this. I thought about how the rock was dull and jagged at the same time. It seemed like a divider of the world, separating water at the bottom from a glimpse of sky. As I pressed the gold leaf onto the cold rocky surface, I felt so close to the cave wall. Close enough to fall through.

In some ways it was hard to deal with that closeness, because I wanted to respect how things worked and what people did in Myanmar, and also because I hadn't seen people interact so closely with wild animals before. Asian elephants have been kept in captivity for millennia and 30 to 40 per cent of living Asian elephants are in captive settings. All working elephants are wild animals kept in captivity rather than domestic animals. They don't meet the scientific criteria for being domesticated, because they don't have successive generations of selective breeding by humans. If there is reproduction in captivity, it happens on the elephants' terms. Historically, some elephants had been reported to mate with wild counterparts in the forest when they were released at night. Could it be they preferred those males to their fellow captive ones, who had to be submissive to their oozies as part of their working lives?

In the face of the complexity, I retreated a little to those scientific and 'objective' facts about elephants with which I felt more comfortable. As with most mammals, elephants have what's known as a U-shaped mortality curve. It basically means that among all individuals born alive (so excluding stillbirths), the risk of dying starts off quite high in the first days of life and then drops before increasing again in old age. Humans have dramatically reduced infant mortality rates and extended adult longevity, so our curve ends up looking more like a tick than a symmetrical U, because the infant mortality drops down rapidly and picks up again very slowly. But remember, these are population-level statistics built on the lives of lots and lots of people or animals. We live in populations, but we are born and die as individuals, and we usually only do each of those things once (or once per lifetime, if you prefer). The likelihood of my or any other animal dying might change because of my age, my sex, and where I live. Sometimes, I envisage myself walking around with that level of risk hanging over my head like the Sword of Damocles, if only a dagger-sized one at the moment, which might one day look more like a bomb with a big red timer, a grand piano, or a comically oversized anvil. As well as the population-level traits predicting my end, I also have my own individual frailties and risks that are part of me, such as those written in my DNA, some of which aren't constantly switched on, but could sometimes be expressed because of my environment or age. So adjusting that more general risk to my very own personal timebomb, if you like. Of course, the sword/bomb/piano/anvil might never drop or explode, because the process could unexpectedly speed up. I could die in an accident, perhaps in one of those tiny planes I keep getting on to see very big elephants, or I could be caught up in a natural disaster, or I could meet one of those grisly ends one sees on crime dramas (blonde women seem to have a very high death rate in those). The effect that such a demise would have on my family might be huge, but in the big picture of adult human mortality in the UK, it wouldn't even create a ripple. This kind of individual variation and the contrast between the idiosyncrasies of a single life and the broad patterns in the life cycles of a population is what made me interested in life history. And in that respect the parallels between humans and elephants abound.

We already know that humans and elephants are social, and they make sense in groups. Families play an essential role in the lives of elephants, particularly for young individuals and for females through their whole lives. There are lots of social species. Apes such as chimpanzees or bonobos might come to mind, but there are many others, from social spider species that build communal webs to naked mole rats – those essentially cold-blooded, saggy-skinned and phallic but also quite toothy mammals with subterranean colonies. What was so exciting to me was that the pace of life in elephants, the timing of life events, feels very human, and that makes it easier to relate to. This idea of breast-feeding exclusively for around the first six months of life, but perhaps continuing beyond that for up to three to five years; the offspring's dependence on the mother being prolonged for many years, adolescence taking place in the teenage years and reproduction after that, in the late teens for females. This was often followed by grandmother-hood for females. I didn't have to look hard to find the similarities with the human trajectory, but it's not because we're closely related. An elephant is closer to the beautiful manatees or that family of hyraxes I knew in Samburu. It's the parallel but on some levels convergent evolution of human and elephant life histories that I found so striking. What does a human environment have in common with that of an elephant that means it makes sense?

I had to start from the beginning and the end, at the same time. The seasonal patterns of deaths and births: I plotted them out, looking for any interesting lumps and bumps in the number of events per month. I wanted to try to explain them. The deaths made sense to me, increasing in the hot and dry months, and also at very cool times. Just as in humans, extremes of the usual climate are linked to higher mortality. I thought about my own experience of extremes of climate. Alternately lolling under the air-conditioning unit and feeling as though I was fixed to the spot outside my flat in the soupy Hong Kong summer humidity, or in Myanmar in the dry season, when we couldn't imagine it ever raining again, and when the Thingyan new year's festivities being celebrated by pouring water on each other seemed incredibly exciting and guilt-inducingly frivolous at the same time. Or the icy grip of a cold winter in Berlin, in which I bundled my dog in a series of jumpers and jackets that rendered her almost immobile. These were just experiences, more illustrative of my ability to travel than of life and death, but examples of extremes. I thought of the familiar news stories of heatwaves and cold snaps heightening the risk of death for more vulnerable individuals, those with reduced immune function, older people or children, and the following population-level increase in deaths. I don't think those patterns are specific to humans or elephants, they're more about being alive. But perhaps it's useful to remember that we're alive just like all the other living things, and that's one of the first things Myanmar elephants made me realise. I am alive, and all the nice houses and hallowed university halls can't protect me from it, nor can packing myself in make-up and pretty clothes or techy active wear. The fragile human form at its heart is still just another flicker of life.

The birth patterns perplexed me more, as I followed the jagged monthly peaks and troughs. Elephants have continuous reproductive cycles, which basically means that, just as humans having menstrual periods throughout the year, the elephants can come into oestrus throughout the year. They aren't like a blue tit or a red deer, which just have an annual breeding season, or a butterfly that bursts forth and only has one breeding season. While those menstrual cycles are around four weeks in humans, elephants have cycles of around sixteen weeks, with only two or three fertile days. My introduction to this wonderful world of elephant reproduction came from an astute and un-assuming scientist from the Smithsonian Institution called Janine Brown. She fascinated me in many ways, not least because she managed to combine being a very deft scientist with having an impressive collection of jewellery (much of it elephant-themed), tattoos and one-liners. On an early field trip to Myanmar, she spun tubes of elephant blood with a hand centrifuge, carefully attached with a vice to a wooden bench. As elephants and oozies wandered in the background and Janine cranked the handle in the shade, the tubes splaying out as though they were on a fairground ride, Janine grinned and said, 'Serum is like liquid gold.' As a lab scientist who could extract traces of hormones from that blood, she was right. But I wanted to mine that demographic dataset. It was my kind of gold.

If the climatic conditions were so crucial in the distribution of elephant deaths, why weren't elephants being born in the peak productivity of the landscape in Myanmar, the monsoon? At that time the rain fell heavily each afternoon in thick fast drops, and the ground smelled nourished all day. The trees would come to life, at first slowly and then in explosions of green. Everything would buzz into action, not least the mosquitoes, and my skin would blossom with huge swollen bite marks. Why didn't elephant lives start then, when everything was bursting forth? It could have been about conception; perhaps the females were conceiving in the life-giving monsoon when their body condition was at its best. But it didn't make sense; with a gestation period of 20 to 22 months the conceptions weren't in the monsoon either. In fact, they were in the dry season, the time of year that I thought would have been, well, barren. What I was potentially ignoring was the other half of elephant life, the life they share with humans. As I knew very well, these elephants had worked in logging; that's why they were so valuable and that was part of the reason Orwell thought twice or more about killing one in *Shooting an Elephant*. And that vital annual rest season, when elephants and oozies have a break from work and are able to interact more with other members of their species, just so happens to be in that dry season. In Western countries, we see a spike in births in September, which is 40 weeks, or a human gestation-length, after the holidays in December. What I saw in elephants was effectively the same thing: around 40 per cent of the annual conceptions were taking place in that three-month rest period. Strangely, I felt ambivalent about it. On the one hand, it makes perfect sense – elephants might have more access to each other in the non-work period and they might also be better able to support pregnancy in terms of physical condition. But this was beyond humans and elephants as reflections of each other. This is reaching through the mirror and affecting one another. On the other hand, it was humans changing the lives of elephants and I didn't know how I felt about that. Elephants look so big and powerful it seemed wrong that we could do that. And it definitely meant that my romantic images of elephants in the forest had more human influence than I had wanted to imagine.

But I couldn't escape romance in Myanmar completely. I stayed on to take a trip to Bagan. It's a place that made up for what it lacked in elephants with many temples. They stud the plain like beautiful and spiritual little hills, most with a large stone Buddha image inside. I think I have been back four or five times since the original trip, but I remember the first time being so peaceful. There were no electric bikes on the roads and no streetlights at night. There was no internet in my hotel so I couldn't see the emails from the editor about my first paper, or be tempted to write one to my boyfriend. I had to live in the time and place. I took a bike out and rode the sandy road to the many temples, or rented a brightly painted horse cart. I did more posing for photographs and noted most of the tourists were from Myanmar. When the New Year festivities started and the sins of the previous year needed to be washed away, I took a number of splashes of water with all the good humour that I could muster, before seeking out the highest temple I could find. I climbed stone exterior steps, each layer steeper than the last, until I reached the highest point. I wasn't alone, for there was a man with his head shaved and in deep maroon robes. He lounged on the corner of the pyramidal summit, looking out over the yellow flatland with all those temple jewels. He smiled. I gestured to my camera, something I always did to ask permission to take a photo, and he nodded. This was my perfect image of Myanmar, capturing the religion, the timelessness, the parched, dry ground all in one vista. It felt too achingly beautiful to be real. I lowered my camera. The monk turned to me. 'Do you know the score in the Manchester United match?' he asked, in flawless English. I stifled a laugh, told him I didn't, and I was sorry. I tried to shake off my assumptions. I couldn't limit this place to what was pretty or what I wanted it to be.

I was able to visit towns and elephants that were at the

heart of timber extraction, like Kawlin and Katha. Katha, the home of the young elephants and their blue track-suited oozies, was particularly special to me. Part of the reason I was so excited to be in Katha was the Orwell connection; he was stationed there in the 1920s as a colonial police officer. As I walked from the compound where the timber employees lived and around the town, the market, the former British Club, the tennis court, the Anglican church, I saw the sights from *Burmese Days*. I wasn't in that novel, or any novel, but I was in a complicated place; as well as the lively modern town, I felt the ghosts. I was lucky to visit what people say was Orwell's house. It was built of dark wooden planks, and although some window panes were smashed and the interior was spartan, it had a charm to it. I've been back to Katha since and once more saw the piles of fruit at the market, the old Singer sewing machines that produce bright embroidered clothes to this day, the sandy path to my favourite restaurant U Soe Thin with benches outside and more chips than you could ever eat. I seem to be chasing the ghosts of my own past: the spirit of 'me' back in 2012 looking across the river before getting on a beautiful boat with a sun-bleached Myanmar flag raised high. The 'me' that waved at those on the riverbank and scoured the wide Irrawaddy for signs of river dolphins, for something to break the surface. But the only thing I saw was my own reflection and the start of the elephant in it.

CHAPTER 4

How do they get so big?

A calf at birth weighs about a bag of rice.
U Toke Gale

'She can come now!'

An older female elephant lurched forward and stepped onto the wooden platform that had been hastily but meticulously constructed hours before. Each of the wooden planks was the same length and thickness, nailed carefully in place. The platform sat in a shallow dugout bed so that the difference between the ground level and the wood was barely noticeable. The elephant stood perfectly still for a moment, and I noticed a cloudiness in her eyes. Her oozie sat on his usual perch atop her head and looked down at me, waiting for my okay to let them move on. In turn, I was waiting for confirmation, from the reader in my hand, for a weight reading in kilogrammes. A series of four figures before and one figure after a decimal point flashed in front of my eyes.

'Two-zero-five-five point five!' I shouted, before looking up at the oozie. 'You can go.' He kicked his legs, and the elephant moved off. She then knelt down, and the oozie climbed down onto her bent leg and jumped to the ground. He gestured to his bag and machete. Should he take them off? I shook my head. No, bring everything you wore on the elephant. He stepped onto the platform and the figures flashed back to life, this time just two before the decimal point. 'Forty-three,' I said, swearing never to step on the platform myself. I pulled down my cap against the sun and smiled at the veterinarians standing next to me.

This is the life, this is science. This is how to weigh an elephant.

The actual process of weighing an elephant is quite easy. Well, weighing a timber elephant in Myanmar is quite easy. I had four thick load bars and connected them via rubber-clad cables to a reader. Of course we wanted to distribute the elephant across the load, hence the wooden platform, and we found the elephants were very uncertain without their oozies in place, so I got into the habit of weighing oozie and elephant together, then having the oozie and any excess weight caused by harnesses, wooden bells or chains go on the scale, and subtracting that from the combined total. The hardest to weigh were the young elephants, who had not been separated from their mothers or assigned their own oozie. Sometimes we managed to coax the calf onto the scale with their mother by waving tamarind and salt treats in front of their trunks and using our most gentle voices. From time to time an elephant with a metal bell came along. By tradition, these bells were fitted to elephants that had killed a person, often an oozie, since they are usually the closest people to any elephant. At those times I stood back and probably secreted all manner of terrified pheromones and made the most concerned of expressions. I was lucky that these didn't evoke a response in the elephants; I was probably buffered by the presence of all of those oozies. Either way, we weighed them all. Sometimes fifty or more a day, hundreds of measurements over the three years that I did fieldwork.

We had a real production line. The elephant had their ID number written on their sides in chalk. They would go through the process of being photographed from several angles by my colleague Alex, a statistical whizz of a scientist who just happens to take the best shots around. It was kind of like a mugshot, but in this case the subject hadn't done anything untoward. They would then have their blood collected, those samples bursting with potential data including DNA, hormone levels, inflammation markers, blood-borne parasites and more. This delicate operation was undertaken by trained veterinarians, the local elephant vets and also team members, who were all able to interact with each other and exchange stories and tips. The whole production line was building up an incredible database of physiological and physical data on top of the demographics from the logbooks of deceased elephants. And in some cases they matched. I was delighted to read that birth heights of baby elephants were recorded in the logbooks of their mothers and each book contained a section for veterinarians to record body height, length of the back and even tail length of that particular elephant. According to the wonderful writer and timber elephant expert U Toke Gale, those tails can grow up to 100 centimetres, and the record length was 155 centimetres. Like U Toke Gale, I've rarely seen such a long one, and oftentimes they can be bitten off or broken in fights, leaving a stubby vestige. The tails that remain intact usually sprout thick and wiry hairs at the end like a brush, each as characteristic as the ears and skin of the elephants to which they belong. Gale wrote that everything, not just size, but body shape, including the shape of the head and the curve of the back, was something to be studied and understood.

While the elephants patiently waited for their turn on the scale, I watched one break a young branch from a tree with her trunk and pop it in her mouth. Things like leaves, grass and bark form the backbones of the elephants' diet. Of course this varies seasonally and geographically, with elephants eating what is available and some having to travel further between patches of food than others. African elephants in the deserts of Namibia don't appear to be particularly genetically differentiated from other savannah elephants, but they certainly have unique behavioural adaptations to be able to cope with the arid conditions, for example by getting water from the stems of plants in the riverbeds. Those elephants are at the extreme, but all elephants have to spend a lot of time eating to maintain their bulk, around 16 to 18 hours a day. That kind of daily schedule means elephants don't spend as much time as us sleeping, only a few hours a night. During their long days, they consume about 10 per cent of their body weight in food (that's over 200 kilogrammes for the elephant at the start of this chapter). In Myanmar, there is enough forest remaining for the timber elephants to forage in that, without the costs of food rising exceptionally high, but for other captive elephants that's just not an option and the associated costs must pile up. I thought again about Orwell likening a timber elephant to an expensive piece of machinery, one that needs upkeep and maintenance and fuel. Of course they are more than that, but their needs are undeniable.

I had always been fascinated by big things, big bodies, big organisms. Much to my regret, I don't live in a great age of megafauna. After the demise of the big dinosaurs, there was a vacancy for large-bodied organisms such as mammals (and, to a lesser extent, birds) to fill this evolutionary gap. For elephants, the peak was undoubtedly the last Ice Age, the Pleistocene, which ended around 11,700 years ago. While I can't ask my grandmother about it, on a geological timescale that's just a mere blink of an eye. So tantalisingly close. The Pleistocene world was a much colder one, drier than now, with water locked up in glaciers and the much lower sea level altering the shape of continents themselves. And across continents, particularly away from the tropics, the wildlife was much bigger than the animals we live alongside today. I like to imagine walking in the frozen tundra among aurochs, mammoths, woolly rhinoceros – animals that sound fantastical and otherworldly but were very much part of the human experience in the past, as evidenced by the beautiful cave artwork depicting them. How long an errant twenty-first-century human would manage walking about in Pleistocene Eurasia is up for debate. I see my icy romantic vision disintegrating the minute I had to face a cave bear or cave lion. It's all over now though: between 80,000 and 40,000 years ago, as the climate and conditions shifted, most animals weighing over 44 kilogrammes outside the tropics died out.

But size really does matter, even now. Think about it: we consider that evolutionary success is about 'fitness', which is more or less the number of descendants an individual has, either directly through reproduction or indirectly from being related. So my nieces and nephews (five, so far) represent my indirect fitness and I'll always have that, even if I don't have any direct descendants and lavish way too much time and attention on a canine that definitely isn't related to me. If reproduction, direct or indirect, is so important, reproducing a lot and early in life seems like a sensible strategy. But this usually means being small so you don't have to invest in growing a huge body instead of, or at the same time as, doing all that reproduction. The whole 'live fast, die young' strategy seems to come with a small body size. Basically, it favours being a mouse over an elephant. There are negatives to staying small and being a mouse, though. It makes you vulnerable to things a bit bigger than you, and you might become prey. But there are likely to be more of you, because you're not so energetically expensive, not using so many resources as if you were elephant-sized. Big things, they are a puzzle, they have all of those benefits of being bigger than most other animals, but this requires a lot of energy. What allows them to grow so big and what limits them? Thermoregulation? Mortality schedules? Food?

I didn't know, and that's why I wanted to study these questions. I find large animals that aren't necessarily big apex predators like lions and tigers really fascinating. Take the elephant, for example. She is a massive creature and she is herbivorous. She eats leaves and bark and grass, not the easiest meal to extract energy from, and yet there she is, over two tonnes. A human is even trickier because it is a predator, it's not an obligate carnivore and it's big (I'm proudly 178 centimetres tall) but not elephantine. Being able to compare humans and other animals and put ourselves into that kind of context is really revealing about how we live. But what is even better than saying elephants are big, is trying to quantify this. I wanted to take measurements of the height, weight, body proportions, chest size and foot size of lots of elephants at different ages to try to get some understanding of how and when they grow. And I also wanted to understand how all this fits with the rest of their lives – their body condition (at its bluntest, how skinny or well padded they are) and patterns of reproduction. How do they eat enough to maintain that bulk? How do they reproduce enough to make evolutionary sense, let alone being successful? And how can that story relate to another big animal that has a similar life pattern, but is a bit different – an animal such as you and me?

Fieldwork is the exciting side of science, but for a lot of the time I looked for the key to making sense of elephants on my computer screen. I checked data collection sheets, and re-checked data on spreadsheets, made models on my laptop and wrote. Not writing like this, but something much more structured, formulaic, focused on answering scientific questions – the kind of writing not many people read, because it's technical and not exactly pacey. But there's a cleanness to it that I know some people must appreciate, and a pattern. I didn't realise it when Virpi interviewed me on that drab February in 2010, but she was pregnant. Her son is about the same age as my PhD. He learned to walk and talk, he went to nursery school. Seeing his life roll out in parallel to my studies was quite intimidating. He made such progress! But my PhD somehow picked itself up and formed something beyond me. Something I could bind in black with gold writing on the cover and submit and publish. I even got to have a tense viva examination in which I gave up any pretensions of politeness and ate all the chocolate biscuits. We have to deal with pressure somehow.

Even after patting myself on the back and forcing my parents to sit through another graduation ceremony, I felt as though I had more data, more raw material than I could handle. I was lucky. In 2014, Virpi introduced me to two brilliant students. Simon, who was a little shy, took to the analyses swimmingly and I liked his student radio slot playing heavy metal. Jennie was smart and sunny and easy to talk to. Simon and I tackled a practical question and a more theory-inspired one. First, how could we use our advantages in being able to weigh and measure elephants to help people who couldn't? We had photographs of all of our elephants with a 55-centimetre bamboo stick which I held in front of them (sometimes nervously). From these, Simon measured in pixels the shoulder height, chest size, body length and foot size of all the elephants. Just as with horses, those variables can be plugged into numerous equations to estimate body weight. This was great news, because the equipment needed to weigh an elephant could be reduced from a pretty heavy set of load bars, a wooden platform, a reader and someone who could convince an elephant to stand on it (and not get killed in the process), to just a camera and a computer. Perhaps people with only camera survey images, triggered by the movements of wild elephants walking past, could estimate the weight of them. Simon found that shoulder height and body length were a bit troublesome, because the elephant really needed to be standing at a right angle with the camera getting a full shot of the side of the body. Capturing this was difficult enough for us to do with the captive elephants and the oozies there to encourage them into position with a tamarind treat. That's quite a lot to expect from a survey image. Similarly, with foot size, the whole weight of the elephant needed to be on it, so it would spread under the weight, and this meant no vegetation could get in the way and that walking elephants, with their weight spread unevenly across four footpads, couldn't be included. It was the chest measurement that was our most reliable best estimate of body weight – this could be popped into an equation and you could see straightaway if your elephant weighed two tonnes or just one.

Then we did one of my favourite things: plotting out raw data and eyeballing it. Two axes, two plots: weight and age, height and age. No surprise that they followed a typical mammalian pattern, increasing most rapidly in the youngest elephants and then tailing off, with most height gain over by the mid-twenties. I looked at adolescence and there was no growth spurt, as we might observe in humans or some primates. The curve was smooth. In females, weight gain followed a similar pattern, but by contrast, male weight gain never really levelled off. Simon tried lots of models to fit it, trying to find an asymptote, and none of them did. That weight was still creeping up when the male elephants were in their fifties. I know what you're thinking, middle-aged weight gain might not be an unfamiliar concept. But I was very interested. There's still a lot of debate as to whether this weight gain might be body-length growth, something we struggled to measure accurately because of the angles the elephants stood at, or whether it is just gain in muscle mass and tusk size. So there was a sex difference for sure, but one that still hides some questions behind those big chests.

With Jennie, we thought about the balance of life again. My friend Carly likes to think of it as a chocolate cake. Basically, she's saying something we all know as biologists and especially as humans: we have limited resources, we only have one chocolate cake and we can't have it and eat it. So whatever we invest in – immunity, maintaining our cells, foraging (and eating), whiling away the hours watching Netflix – it's all eating into the cake, even if it doesn't seem that way. Jennie and I were thinking of two particular slices, growth and reproduction, and how they might trade-off with each other. When biologists measure reproductive investment (or cake slices), we often think of females, simply because it's much easier. We can measure how much they invest in reproduction because we can, as a minimum, count their offspring, whereas for male elephants we'd need DNA to check paternity. So we focused initially on females for practical reasons. We talked about it and thought perhaps early reproducing females, say those who gave birth before 18, might be smaller, as they would have to limit investment in growth and funnel it into growing and supporting their calf. On the other hand, perhaps they would be larger, because they needed to have a certain weight to maintain pregnancy and gestation. What we found was, as usual, a little more complicated. When elephants were in their peak reproductive years, which was early in this population, hitting the maximum at just 19 years old, taller females were more likely to have reproduced. But this effect diminished with age. So, potentially having a height advantage early on might help elephants to reproduce, but maybe not later on. Jennie found a lot of other things: that we couldn't find such an effect for weight, that body size didn't seem to be linked with survival, that there wasn't any significant link between the size of a female in terms of her weight or height and how many calves she had over her lifetime. Now this is just in the elephants for which we had data, around 100. But I do think that it shows it's so difficult to untangle all of these effects and associations, between size and reproduction, much harder even than trying to share out chocolate cake. I suppose that's why some scientists do carefully controlled experiments rather than relying on data generated by the messy real world.

I thought again about Amboseli, that gold standard of long-term elephant studies and how a different kind of elephant grew in the shadow of Kilimanjaro. It's a different species as well as a very different place, with the elephants ranging oozie-less under the African skies. The Amboseli team found that female elephant hindfoot length didn't stop increasing with age, so potentially they were growing through their whole lives. And they also found that females who reproduced early (aged 12 or younger) tended to have bigger feet than those that reproduced after that age. This isn't just a foot thing; the foot length is a proxy for body size overall. Interestingly, elephants born in dry or drought years also measured smaller, with that foot size and their height reduced, a reminder of the effects of early experience rippling through life. I think about this every time I look at an elephant foot-print in the sand, or the mud, by a river or along a road (elephants sometimes like to walk along roads, as there are fewer trees in the way). In the absence of sightings, those footprints are indicators of the size of the elephants; you could also measure the circumference of one of those dung balls between the footprints.

So next time you're on safari and your guide says it's safe, ask her to pick up an elephant dung ball for you. You can measure all the way around the circumference or just measure one of the ends, take the longer and shorter meas-urements and average them to get the diameter. If that diameter is less than 12 centimetres, it's likely to be a young elephant, aged under 15 years or so, while 10 centimetres or less would be a calf around the age of 10. Take a look as well at the composition of the dung, and you'll notice it doesn't look very processed; that's normal, elephants are hindgut digesters, not ruminant like cows, so there's plenty of visible plant content still there. But even that varies with age. Elephant tusks are super-long incisors and the rest of their teeth are premolars and then molars that kind of move through on a conveyer belt along the gums, wearing away over time ahead of a new set coming through. So as a human, my teeth push up through my gums. If I were an elephant, they would come through the back and move along through. African savannah elephants have the Latin name *Loxodonta africana*, which refers to the sloping ridges of their teeth, not the grand incisors at all. If you ever have to tell apart an African and Asian elephant by their teeth, you can even put the tusks aside, because molars could help you more. Lay the two teeth side by side and take a look at those ridges and ripples. For the African elephant tooth, the ridges are like diamonds, whereas the Asian elephant ones would look more like parallel lines, maybe just meeting, clustered in together like folds or ripples on a beach. Elephants have up to six of these molar sets in a lifetime, and as the last set wears down, the elephants can't grind up their food so well before swallowing it. They do things like carefully knocking the dust off the grass they eat and choosing softer foods as their last set of molars wears down. So, in some of those big dung balls, you might find sticks and a lot of undigested material, a sign of age and a reminder of the circle of elephant life. You might also find a frog, which is a reminder that frogs sometimes live in dung balls.

Most humans weigh less than elephants do (you are welcome to quote me on that). Elephants can be around 100 kilogrammes and 1 metre high at birth. What strikes me is that the total mass of humans on this planet is much, much larger than the elephant mass. Perhaps I am fooling myself thinking that I'm an elephant, because the humans have already won. Maybe my human side will always beat my elephant side. I know that three elephant species survived the Quaternary extinction at the end of the last ice age. Species that lived primarily in the tropics and didn't experience such a dramatic change in their environment and the human occupation of that environment in one go.

But I can't deny that their persistence has meant facing a world increasingly dominated by human occupation, human biomass and the biomass of domesticated species. Just 6,000 years ago, the range of Asian elephants swept from Mesopotamia in the west, through India and southeast Asia, as far south as Borneo and as far north as the Yangtze river. The current range is now patchily strung across the Indian subcontinent and southeast Asia, not even a shadow of what it once was. In Africa, elephants were distributed through much of the continent, from the Mediterranean coast in the north to the southern shores bordering the Indian and South Atlantic oceans, with forest elephants inhabiting the Congo Basin area and coastal areas of west Africa. Estimates suggest there were as many as 26 million elephants on the African continent at the end of the nineteenth century. Today, there are around 415,000 elephants. Their ranges are shrunken and fragmented; what once would have looked like a blanket across Africa is now in tatters, with many areas completely devoid of elephants.

It can be quite conflicting, knowing that I'm only able to study elephants because I'm a human trying to understand how to be one, and yet it's humans that are their biggest threat. My humanity is a threat to them, and somehow they still exist, even coexist with us. They still touch my hand with their trunk and wrap it around, letting me feel the rings of skin and the spiky hair, and the trunk tip, sticky with mucus. Under the burden of such information, it's easy to lose yourself in a small and repetitive task, like weighing elephants. Just have one after another walk across the scale and soon you'll drift away. It's more effective than counting sheep, that endless parade of elephants; you can fool yourself that they are plentiful and common, that you're the one outnumbered, that the world is more elephant than human. But it's not. That's why the big picture matters.

You're never 'just' weighing an elephant. As scientists, we have the big picture in our minds all of the time. How does the data point of body size fit into our understanding of the elephant – on an individual level – if this animal is smaller or bigger than we'd expect? Has it been ill, given birth, lost a mother or experienced something else that could affect it? At the population level, how does this animal compare to the others? What's the range and distribution of sizes? And then, at the evolutionary level, what are the benefits of this size? Is it that the only real predators for adult elephants are humans? How do they compensate for delaying reproduction for so long? Is it those family ties and breeding together over a long life? It's not easy to put the puzzle pieces together for such basic bits of data. We need to, though, because the answers don't just satisfy intellectual curiosity, they tell us about what the conditions are for a sustainable population; how traits like body size might shift under different mortality schedules; and, you can probably see what I'm getting at here, how to keep elephants alive.

According to U Toke Gale, the skin tone and mottling that Asian elephants acquire as they age is important and gives clues to their temperament and working abilities. I've never seen it on African elephants, but as they age Asian elephants start to have pale skin with freckles of darker pigmentation. It's undeniably beautiful and an appealingly attractive trait to look out for, unique like a fingerprint. But those irregularities in pigmentation don't make them a white elephant. There are critical criteria to achieve that label. I didn't need those the first time I saw a white elephant. I just knew. They were pinkish, not white, but I never could have mistaken it. The lack of pigment is often not caused by albinism, which means a complete lack of melanin and gives the animal pink eyes, but leucism, a partial loss of pigmentation. Either way, white elephants have long been a symbol of prestige and political power in the majority Buddhist countries of southeast Asia. The Buddha entered the womb of his mother Queen Maya in the form of a white elephant in a dream. Outside of dreams, they sometimes, very rarely, occur in the wild, for example, several have been caught in the Rakhine Yoma area of Myanmar, a currently unstable region next to the border with Bangladesh, where they are so valuable that they were and continue to be kept in special compounds. U Toke Gale wrote vividly of a male white elephant that was separated from a large herd in 1806. It was transported by barge to the then capital of Amarapura, where it lived in its own palace, decorated with mirrored glass and overlooking the royal palace itself. Elephant reflections in tiny pieces of glass.

This level of celebration and respect doesn't surprise me. I've been to the modern capital of Myanmar, Nay Pyi Taw and I've seen the empty streets punctuated by roundabouts bursting with plants in the arid climate, an enormous temple (a near replica of the Shwedagon in Yangon) and the white elephant enclosure opposite. Even though the country is now a republic, the symbolic sway of the elephants has not faded. When I last visited, there was an adult female and two young white elephants; the youngest was her calf, which had been born after she was captured. The first time I met the white calf, he played with a tyre and a ball, and it was charming and delightful. The next time, just a year later, he had grown enormously and was now powerful as well as playful. Banana treats had fuelled his growth, and I hoped he and his keepers would be safe and able to contain his youthful enthusiasm. All three white elephants reminded me of mammoths, their hair thick and falling in unruly blonde strands. I put my hand on the side of the oldest female, the mother; her skin was pinkish in tone, just like me, but a little harder. Her eyes were bluer than mine, pearly and almost iridescent. Somewhere between cornflowers on a sunny day, and a flash of lapis lazuli of a kingfisher darting into a river. They changed in colour and opaqueness with the light. I rested my hand on her and noticed it was blonde on blonde. My eyes, grey-blue and unclear, like worked glass, held the clear gaze of the elephant. I didn't know what to make of her, I think it was beyond me. I didn't know her name, so I whispered to her, *Sin phyu daw*, royal white elephant. I touched her trunk again.

I noted that two of the first words I learned in Myanmar were *shwe* (gold) and *phyu* (white), like those elephants. On one trip, we gave up on the gut-shaking minivan of shades of brown and took the train from Mandalay north to Kawlin and Katha. We bumped along the tracks, accumulating lateness gradually, like sheets of time slipping across each other much more smoothly than the carriages on the track, until we were later than the trip was long. I let go of my understanding of time and sat in the dining carriage with Alex, Aung Thura Soe, an elephant veterinarian who could by turns be commanding and fun, and a researcher named Adam with a ginger beard and a circumspect streak. We drank beer from chipped mugs painted with English flowers. Another man sat with us, an Australian. We asked where he was going. Myitkyina, he said, the end of the line. He wanted to get away from all of the Western people. I wondered how many stops he'd have to go to leave himself behind. I watched the villages, white temples with gold stupas and bullock-ploughed fields pass, at a speed barely above a jog. Aung looked and me and smiled. He told, you're white and gold, that's why people like you here. I giggled uncomfortable. What else could I do? My feelings of guilt and inadequacy, undeserving of any special attention, were just another indulgence, privilege, dare I say it, I got from being white and gold.

I had not come here to think about my place in the world, but it's not possible to be there without noticing it. I was the only woman in that train carriage, with the men and the booze. Off and on the train, in the streets, restaurants, temples and while we worked, people stopped me, Virpi, and our Finnish colleagues every day for photographs, as though we were special. I held babies, I held hands with girls, had people touch my hair and say it was just like a spider web, I accepted flowers and grinned awkwardly over and over. In Chin State, I practically destroyed an outdoor loo that was essentially a closet on stilts made of wooden slats. It just wasn't built to accommodate someone almost six feet tall and yet the owner beamed with pride that I had visited.

Our train slowed even further and creaked to a halt, the message passed down the carriages that the train ahead had derailed and we had to wait for hours by the tracks. We eventually made it upcountry to see the elephants. The train achingly rousing back to life and rolling on through the dark. As we prepared to leave, and collected our numerous bags and supplies, a man caught my eye and whispered to me, 'Just look forward, don't look back at me'. I swallowed and nodded. He put something in my hand. I wasn't afraid of him as I walked past to jump off the train. As I stood on the platform, I opened my hand and found a white plastic pin. It had a fuzzy colour image of the activist and pro-democracy leader Aung San Suu Kyi in the centre. Around the edge, there was a slogan in bold English lettering: We Must Win. Then, shortly after her release from house arrest and before she was able to stand for elections, it felt like resistance. Later, with the sweeping victory for her party in 2015, it was what people had fought for coming to pass. And now, with the upheaval and military interventions in Rakhine, and her star fading internationally, it's hope again, that the 'we' is more than the majority. A knowledge that who is good and bad and what winning means is too complex to print on a pin. I never saw that man again. Back then, it had been a time to be very careful with such displays of allegiance. Things shifted over time and I saw Obama's motor-cade visit, the British Prime Minister also came and then the investors, followed by the proliferation of non-profits and the opening of the country to the world. I felt slightly superior at having arrived first. But I was really there for something else. I was there for the elephants.

Even elephants were always more than I initially thought them to be. We know elephants for their size; there's a reason that the name of one ill-fated elephant, Jumbo, became the byword for anything big. His was a life in which humans transported him from Sudan to Paris to London and then to North America. He was an exhibit in an age of discovery, a symbol of the scramble for the African continent, a pillar of strength (and one that was often overloaded, evidenced by the enlarged tendons in his back) in the industrial age and a star making it in the New World. He walked over Brooklyn Bridge to demonstrate its safety, attracted crowds in Paris and London and then in Times Square. But it ended tragically. Jumbo was hit by a locomotive, another powerful engine of the industrial age, cutting short his life at the age of just 24. His death in St Thomas, Ontario, was miles from the range of his species, but his mythology was bigger and had more longevity than his physical presence. And that's saying something because he was huge, measuring 3.2 metres at the shoulder. Perhaps if his life hadn't ended abruptly, he might have reached the 4 metres that showman and circus owner P. T. Barnum claimed as his size. There's something terribly sad about a huge elephant showing weakness, and I can only imagine what Jumbo's long-time handler Matthew Scott felt as Jumbo lay dying. We still have his skeleton, the end of his tail, and the glimmer of his stardom, like the echoes of a distant elephant trumpet.

The biggest elephant I measured in Myanmar was a tall and beautiful male with thick tusks called Thaung Sein Win. We're not meant to have favourites, but I know several people in Virpi's team had a soft spot for him, as he elicited mentions in thesis acknowledgements and social media – less of a star than Jumbo, but a local celebrity for sure. He had the sort of amber eyes which some elephants have that I find quite piercing, and he was a very olfactory elephant, always sniffing with his trunk and holding it to the roof of his mouth, where elephants have a sensory organ to process scents. He was undoubtedly grand and incredibly strong. He was our giant: 3.6 metres tall and weighing in at 4 tonnes, with a speckled face and beautiful thick white tusks. Thaung Sein Win was already about fifty when I met him – only he really knew exactly how old he was because he wasn't born in captivity. Like Jumbo, he started his life wild, living with his mother and a herd. When he was only about four years old, they were driven into a stockade (sometimes known by the Hindi name 'khedda') and the group became timber elephants. This method of capture was frequently employed in the past, having the benefit of acquiring a large number of elephants in one go. It must have been quite a spectacle, however difficult it is to contemplate now because handling and welfare practices have changed so much. Oozies trained the elephants, as well as people on the ground who would herd the elephants into a cleared area. This would then be fenced in, and fire, drums and a general din were used to funnel the elephants into an enclosure. I should say that Thaung Sein Win is around the age of my father and I didn't witness his capture, but it took place in the area around Myitkyina, in Kachin State, in the heavily forested northernmost reaches of Myanmar, the tip of the 'kite' that juts into China. The process by which he was captured is no longer used, forming just part of the history of elephants and humans in Myanmar. In fact, elephants are now recognised as endangered and therefore not routinely captured to become logging elephants at all. But Thaung Sein Win's generation is a connection to a not so distant past, one that stretches back to the elephant captures that the ancient Greek historian Megasthenes witnessed as an ambassador to India in antiquity, in which his descriptions indicate a stockade not too different from those in the nineteenth and twentieth centuries.

Thaung Sein Win grew to leave his mother and the elephants he was caught with, to be moved to Kawlin, a town on the railway line that snakes north of Mandalay. It's a place that is pretty unexceptional visually. The market can be a little dreary and there's a large golf course, a former plaything for the Brits, and, like every town or village, a temple that I have been to many times, paying particular attention to douse the lion with water. (The lion was the symbol of the day I was born on; if you were born on a Wednesday, you were in luck and your sign was the elephant.) But the place will always have the transient beauty of a summer romance for me. When I first went to Kawlin in November 2012, it was bathed in golden sunshine and home to some incredible people working with elephants. There are few places where I have been happier. And the most special thing is that you can work all day under a green plastic shade, processing your faecal samples and your photographs and measurements. Then in the softer early evening light, if you look up you can see Thaung Sein Win standing tall, incapable of not looking majestic. In that moment, you can believe there really are giants in this world.

CHAPTER 5

Sex and baby Hannah

their massive blood
moves as the moon-tides, near, more near
till they touch in flood.

                                    D. H. Lawrence

I once learned how to do an ultrasound on an elephant. I was in Thailand, before I even made it to Myanmar for the first time, before I got my PhD, had my first mohinga (life-affirming noodle soup) or questioned whether I really liked the Beatles or it was all about conditioning. I was now in Lampang with a group of students and teachers. And there were elephants with their mahouts (the word more commonly used in Thailand for the equivalent of an oozie). There were some very kind veterinarians: the patient Chatchote Thitaram and his team, and Nikorn Thongtip, the eternally youthful and skilful research veterinarian and academic (who I can officially vouch for as the most fun person to win a cash prize from a competition on the back of a beer top). Chatchote showed us the portable ultrasound, the screen and the sensor. We all knew elephants were pachyderms, so their thick skin would prohibit an external examination. This, my friends, was going to be transrectal: going in through the anus.

One of the assistants showed us how to turn our shoulder-length gloves inside out so that the seam wouldn't hurt the elephant, and we taped the tops of our gloves to our scrubs. We put on aprons and tied back our hair. The assistant then ensured the path was clear for the ultrasound by gently removing the dung boli from the elephant's passage. We stuck a thermometer in the steamy dung to estimate the body temperature of the animal. I watched the assistant put paraffin for lubrication on Chatchote's arm and he was ready to go, wrapping his hand around the sensor to ensure minimal discomfort to the elephant and easing his arm into position. He showed us the screen, which we shielded with a sheet to block out the glare; there was the bladder, and there, the seminal vesicles (cigar-shaped organs where seminal fluid is stored, and these were full) and the ampullae (cone-shaped structures where sperm are stored, full too). It was a male elephant, and it looked ready. But I was not.

A few years before, the famed veterinarian and elephant reproduction specialist Thomas Hildebrandt had visited Thailand. He trained Chatchote and Nikorn and the team with me that day in Lampang on how to collect some critical samples. Semen samples. One half of the beginning of an elephant life, or at least the potential for a beginning, with a long way to travel to get there. The female reproductive tract is no piece of cake to traverse; there's a whole vestibule (an additional tract about 1 metre long) that the sperm need to traverse before even reaching the vagina. Hildebrandt had developed a technique to acquire semen samples from elephants that was reliable and replicable. Essentially, it was a manual operation. Nikorn and two other veterinarians showed us in Lampang. We had already done the prep: clearing the tract and checking the ultrasound. Nikorn had made tubes to collect the samples on long poles. He took the net off the end of a fishing pole and in its place put a plastic tube surrounded by a bag, hanging over it. Nikorn and the other vets showed us exactly where to put pressure, using the ultrasound, and exactly which movement to make. You put your hand into a fist and push through from your wrist, rhythmically. This was a precise operation with one person doing the movement and two people with the tubes on fishing poles ready to collect the sample, while at the other end of the animal, the mahout calmed and talked gently to the elephant, stroking his trunk. I didn't know what to expect. But it was successful and productive, and within ten minutes we had three tubes of semen, not contaminated by urine at all.

Nikorn suspended some of the sample in egg yolk and carefully pipetted a few drops onto a microscope slide. He ushered me and some other students over to the microscope. And there they were, elephant sperm, looking more or less like any other sperm I had seen: head like a drop, long tails, lots of movement. There were a few with two heads or short tails, but nothing spectacular. But all of this not-so-special wriggling entrapped in vitellus was actually a bit special. You see, not all elephants reproduce in captivity. We don't know why, it could be so many factors: proximate ones like hormonal issues, reproductive tract problems in females, male semen quality, or wider ones like access to other elephants, diet and behaviour and dominance. In the past, when elephants like Thaung Sein Win were caught from the forests and Jumbo was caught in the Sahel, the supply of elephants seemed endless. We can't fool ourselves anymore that this is possibly the case. What's more, elephants are beyond being an economic resource to most people who interact with them in captivity. Zoological gardens are more likely to frame having a limited number of elephants or other threatened species as important to serve as an educational tool, to pique the interest of visitors in conservation and, in the worst cases, to provide a pool of genetic material if the wild populations dwindle to such a point that low diversity (inbreeding) could prevent the populations from ever recovering. The very fact that elephants are in captivity at all is a talking point, and understanding how elephants reproduce is vitally important because there is huge pressure to make the population sustainable.

What we learned went beyond just acquiring semen samples, but also included carefully monitoring female reproductive cycles to determine fertility. We do this in the laboratory, with hormones isolated from blood samples, but you can also ask an experienced male elephant. Chatchote illustrated the latter point to us on our training trip with a simple sniff test. He had urine samples, in trays (as there's so much liquid every time an elephant pees), from a female in oestrus, one close to oestrus, but not fertile, and one that was miles off. He presented the samples to an older male, one that had successfully mated with a female before. The male hovered his trunk over each tray he was presented with, breathing it in. But the oestrus sample brought a strong reaction, something we call the flehmen response. Sadly, it's not the same as in ungulates (mammals with hooves), when they curl back their lips and show their teeth. I can't imagine an elephant making that expression. He did the elephant version: he dipped the tip of his trunk in the urine and then lifted it up and touched the roof of his mouth with it. Have you rubbed your tongue along the roof of your mouth? Mine's quite smooth. If you were an elephant, you'd have orifices there that lead to something called a vomeronasal organ, which is a smelling organ that they use to detect chemical signals, and that's why the male did the flehmen so enthusiastically with the oestrus urine. It had a distinctive olfactory signature that he could pick up. When Chatchote repeated the test with a younger male, with no sexual experience, he was doing the flehmen with the oestrus urine, but also that taken from the female around the time of oestrus. His detection just wasn't so sensitive, and the reaction was somewhat excitable. That youthful enthusiasm made me smile.

While oestrus is essential to female fertility, males experience a phenomenon known as 'musth'. This has been known about for centuries. I've always smelled musth before I saw it, and I have nothing like the olfactory capabilities of an elephant. It smells like a farmyard. Now what's making that smell is the sticky secretion from the elephant's swollen temporal gland. He is also likely to have his penis erect, and dripping with urine; you'll often see urine stains on his back legs. While these physical signs are certainly there, what can be just as apparent is that the elephant seems agitated: he might tusk the ground, he might fight with other males, sometimes to the death, follow family groups and generally be more unpredictable and aggressive than usual. He's likely to forage less because those other activities take up his time. He might emit a specific kind of low, metallic pulsating vocalisation called a musth rumble. Because of all of these unusual behaviours, such elephants are hard to predict. For the sake of your life, it's best to avoid a male elephant in musth, whether he's African or Asian.

In fact, on the day I was doing ultrasound with Nikorn and Chatchote, there was a captive male elephant in musth in the area. He was far enough away that I didn't even get a waft of his scent. He was tied to a tree and looked relatively calm. But the bulging of his temporals and the urine on his legs indicated that he was best avoided. Biologically, beneath the stench and the behaviour, what he was experiencing was a huge surge of androgens: male hormones. This is a little different from human males, who are quite consistent in their production of testosterone and other sex hormones, although these can decline with age. What I still don't think elephant researchers are in agreement about is how much musth actually has to do with sex. Studies have shown that males in musth can beat out competitors in tussles for dominance, and that they're interested in following females and mating if possible. But it's not necessary to be in musth for a male to mate, and all adult males are interested in oestrus females. Musth might not be just about sex, but also about determining position in male society and communicating body condition to other males – the same individuals who are their potential competitors. The male I looked at gingerly in Thailand that day appeared in decent shape: his face was not concave, and his spine wasn't standing out like a very steep mountain ridge. He was probably only at the start of his musth cycle, because eventually all of the not-eating and doing other things causes a drop in body condition, and with it the hormones drift down too. Some males maintain musth for a few weeks, others a couple of months, and the majority of older males establish an annual cycle. But like most things I was learning about elephants, it was more complicated and variable than I had initially imagined.

Sometimes, when I was learning about anatomy and physiology, it seemed incredible that there were any elephant calves at all. Dr Hildebrandt, the originator of the semen collection technique I was lucky to learn, is responsible for elephant calves across the world coming into existence, many of them through artificial insemination. The Thai team had also been able to produce their own AI calf (artificial insemination, not artificial intelligence). Following the semen collection and careful monitoring of female reproductive cycles (based on blood tests, not easily aroused males), the other step to bringing an elephant into being was the actual insemination. It might sound straightforward; farmers do it with their livestock all of the time with reliable results. But remember this is elephant scale and this is wacky elephant physiology: the opening of the vagina is at the end of that 1-metre vestibule, which frankly sounds like an architectural term for a reason. It's all titanic stuff, but requiring precision timing, knowledge of anatomy to a particularly fine scale and being steady-handed around animals that could definitely kill you if they just decided to kick you or sit down on top of you. It's a testament both to science and those incredible individuals that these calves are made.

Whether conceived the tech-heavy way, or the old school way, on Chatchote's ultrasound, you can also see an elephant foetus at certain stages of development. They have trunks! I don't know why I found that so exciting. I mean, I used to have a tail, so it's completely logical they should have append- ages that they still possess post-utero, whereas it is with a little wistfulness that I confess my tail disappeared in utero. It was a vestige of a different me from a different time. But that tiny noodle trunk, it breaks your heart. It also makes me grin at everyone. I can't get tired of seeing that: a little big life.

That course was the first time I thought my understanding of elephants could go beyond numbers or measurements from a scale. I've always been incredibly impractical, but what Nikorn and Chatchote and the others taught me was the kind of things they do that aren't just about getting incredible data. The process itself is also fascinating and it makes you feel connected in ways you could never imagine. The 'doing' of science is as meaningful as the results of science. I saw scientists doing something they thought was important in such a sensitive and skilful and professional way. That kind of thoughtfulness is a service to the elephants and mahouts.

Even if so many elephants in captivity were conceived with this extra help, it's not the typical elephant way. There's a poem by D. H. Lawrence about elephant mating. It captures the grandeur and the silence and the deliberate easiness with which elephants often seem to live their lives. As a text, it's beautifully paced, leisurely, like an elephant heartbeat. But it is not entirely biologically accurate (something I don't think art should aspire to, or at least wear too heavily), giving elephants undue credit for their discretion in the act: a secrecy, a cover of darkness, a co-sleeping. I've seen elephants in the open, in the day, going right ahead, with the presence of a vehicle and curious scientists posing not the slightest deterrent. It's one of those times that we want elephants to be like us because we see that kinship and we respect them, but we just have to remember that elevating them creates distance too and that reproduction and all it entails is perfectly natural. That includes males trailing around after oestrus females, having erections and mounting each other, even with a blonde scientist watching through her binoculars.

Most of the time, though, my observation of reproduction was still in those little green logbooks from Myanmar. Baby elephants were first recorded in the logbooks of their mothers. The elephants officially got their names when they got their own book and were trained at the age of five. Before then, they might have a nickname. I started to learn common components of the Myanmar names, which were usually made of two or three discrete units, similar to the names humans have in Myanmar. The names are themed according to their training year, which can be quite annoying if you're trying to work out if Aung Mya Moe is a different elephant to Aung May Moe, or a typo. I know Aung means 'successful' and Moe means 'rain' or 'sky', Mya is 'emerald', and May is 'girl' or, more archaically, 'maiden'. I liked the naming system, not that I know how elephants think of each other, but just for the human side of it. I'm a firm believer in not naming oneself, but letting other people name you. And I have the most names for the ones I like most; ask my sister or my dog. In Thailand, I picked up the nickname Noona, a little mouse. I think it was intended ironically. In Myanmar, I'm called Byae by the people I like best. It means egret, and I was told it's because I am 'long and white and always around the elephants'. I wear it with pride. But my proudest name isn't even one given to me. It's the name of a baby elephant.

In Kawlin, in 2014, a female elephant was heavily pregnant,

over 20 months. She was shifting uncomfortably and pacing. This is how it happens. Usually at night. Other females are in the area to keep her calm. This wasn't the first time for this elephant; she'd had seven calves in the past, all of which had survived, and she was nearly 47 years old. She was already a grandmother herself. She had felt the pressure and discomfort before. So when the final contraction led to a sack of fluid and blood and life bursting forth from between her back legs, it was something she was less afraid of than she had been the first time. Maybe she was less likely to panic and perhaps kick out in her confusion; but more likely to remove the membrane, touch the contents with her trunk, prod it and encourage it to stand. And stand it did, a baby elephant, a female. She was born in captivity like her mother and grandmother. One that latched and suckled and, over time, over years, grew strong and bold. The veterinarians named her Hannah.

I'm not the only person lucky enough to have an elephant named after them. There's a special elephant called Nyein Carly, named after my friend and colleague, Carly Lynsdale. Nyein Carly's mother was also fecund; she was the youngest of ten calves. In elephants, twins are even rarer than in humans, so that meant Nyein Carly's mother went through pregnancies, of 20 to 22 months, ten times. No wonder elephants live for so long. Nyein Carly is an active and confident elephant. And human Carly probably fits those descriptors too. She's also a committed and fastidious researcher with a very high disgust threshold. Virpi had decided to expand the team and get another PhD student in. I was coming to the end of my doctoral research and helped Virpi sift through the candidates, draw up a shortlist and, the fun part, do the interviews. Carly had the right combination of grit, calm under pressure and commitment to the methodology that we knew would work in a project with a lot of fieldwork. I'm not going to skirt around the issue: Carly's job involved going through a lot of elephant dung. It's no secret that I am entranced by the scientific possibilities of elephant dung and that it's full of DNA ready for extraction (plant matter from the diet, bacterial DNA from the gut microbiome). Equally, I am fascinated by the practical applications – burn it as a mosquito repellent; add chilli to that fire and you've got an elephant repellent; use it as a building material or fertiliser. Carly was looking for something specific: something visceral, something deadly that really makes you want to wash your hands and never eat with them again. That something was worms. Intestinal parasites, mainly.

Virpi was filled with excitement one day when we broke open a dung ball and saw some white worms of just a centimetre or so squiggling around. I, on the other hand, still have nightmares that they hid under my fingernails and I ingested them (and I write this having had the stool tests and medical examinations that should assuage such fears). The excitement wasn't just because we're biologists and we like seeing any living thing. It's more because parasites have been so crucial in shaping the lives of all animals, including big ones like elephants and us too. It links to the chocolate cake analogy that Carly likes to use. If an animal has to invest in fighting off parasites, there's less of the cake to slice up for growth, reproduction and other functions. One of the first things Carly did was to lock down a way to sample from elephant dung. In a lot of species, you can go through all the faeces produced in a defecation event, but elephants can produce five boli, each perhaps 20 centimetres across, every time they go, and they can go seventeen times a day. That would be a real test for my disgust threshold if there were white worms in all of them. In reality, though, Carly found eggs more often than adult worms. I spent some time diluting dung in saltwater and pipetting it onto slides, so that Carly could do the hard work of finding and systematically counting the eggs. Given my laboratory work aversion and frustration every time I trapped an air bubble in the slide, you can guess that I either thought the results might be particularly interesting or I really like Carly. In actuality, both are the case.

Fortunately, Carly found that eggs don't seem to clump in elephant dung balls, nor was she more likely to find them in the centre or closer to the edge. So she saved herself some work by being able to sample. Then she could start thinking about the evolutionary implications of those eggs. As with most host species, the distribution of the parasites she found across individuals was really skewed: many individuals had no or just a few eggs in their dung, but there were a few individuals with hundreds of eggs in there. She found that the younger elephants were more likely to have an infection, probably because of their developing immune systems.

The same was true for older elephants, pointing to declines in immune function with age. And there was also a sex difference. We know that male sex hormones can have the effect of suppressing immune function, so could indirectly cause higher mortality, not to mention the behavioural impacts of high sex hormones, just like those male elephants in musth, which don't eat much and can fight to the death. Balancing the potential advantages and disadvantages is a fine line for elephants to walk. It's perhaps one of those very individual ones. For some, investing in getting pumped full of testosterone might be a smart strategy, perhaps with immediate reproductive gains, but with the alternative, lower testosterone, maybe fewer worms in the gut could lead to a longer life and more opportunities to reproduce over time. It's a balancing act that all species have to perform; it just seems a bit more impressive to imagine an elephant doing it.

Famously, ornaments that some animals have, such as elaborate tail feathers on birds of paradise, have been interpreted as signs that they're in great condition and can therefore invest in such things. Which is sexy, isn't it? Parasites, or having a lot of them, can throw a spanner in the works by limiting the energy animals have to put into such fitness flags. Something worth investigating is if male elephants might do something similar in terms of their tusk size – perhaps it could be negatively associated with gut parasite burden. Parasites aren't just linked to sex, but also to death. They go for the big hitters. Carly found that in the Myanmar timber elephant dataset, parasites of some form are a frequent cause of death; among them is a generic worms category, the disease filariasis caused by infection with roundworms and bots. In case you were curious, bots are a kind of myiasis, which is an infection in which fly eggs get into the host and several stages of larvae grow in the tissue of the stomach. Mature larvae then exit through the elephant's mouth. Eggs can often be found around the base of the tusks. The reason we know these causes of death is because of those incredible records kept by the veterinarians in Myanmar. On most elephants, necropsies are performed so that the veterinarians can ascertain a specific cause of death, and this is rarely possible for multiple generations of elephants in an elephant range country. The logbooks are the goldmine that keeps giving, but the people and elephants behind them are the real treasure.

That was all in the future for Baby Hannah, but to start with, her life was a lot about milk. When I read Aristotle's description of elephants, I noticed he was pretty taken with the trunk. He correctly points out that calves like Hannah suckle with their mouths rather than their trunks. In fact, given that the trunk is a kind of nose and upper lip extension, elephants don't eat or drink through it at all. If you see them sucking up water, they store it there and later blast it into their mouths. A suckling calf like Hannah lifts up the trunk to drink; one of the first things she does is learn to control it. Milk itself is a pretty hot topic for people interested in elephants. Like us, and like all mammals, infants are dependent on it. Elephants of all species are particularly sensitive to other milk types. If people, with the best of intentions, find young elephants separated from their families and give them cow's milk, this can be extremely dangerous, causing diarrhoea, dehydration and sometimes even death. Unfortunately, situations do arise in which elephants don't have access to their families. Colleagues of mine in South Africa were called to take care of a young elephant rescued from a drain at a copper mine. Attempts were made to introduce him to herds in the area, but he was rejected. He was taken into the care of the Hoedspruit Endangered Species Centre and they named him Amanzi (isiZulu for water). Videos of him being heroically lifted from the well and breaking down species boundaries by hanging out with a sheep went viral. People were charmed by this fragile life, and his perking up whenever he was around water. Yet even with the very best veterinary care and knowledge of bottle feeding an elephant, and people across the world offering all of their support and goodwill, Amanzi unfortunately didn't make it.

Humans affect the lives of elephants so much, and in cases where a calf is separated from its mother because of a human structure, poaching or other human activities, I do think there is a moral imperative to intervene. However, it's always the case that a certain number of calves will not make it to weaning. In the Myanmar elephants this figure can be around 15 per cent; a sturdy female I knew in Katha called Pan Cho Lwin illustrates this. She was born in 1977 and had four calves. When I met her in 2012 she had one at heel, following her everywhere, and two older daughters that had grown up and become adult elephants. Years before, she'd had a male calf that had died, aged eight months. The risks for calves in Myanmar vary because of individual characteristics, such as how many parasites they have, but their family structures or their experiences of capture.

Even though young elephants can walk by themselves and don't need to be carried, don't mistake the fact that they're on four feet for independence. Another of Virpi's protégées and now a successful researcher in her own right, the always fastidious Mirkka Lahdenperä, analysed the impact of the presence of other family members on the lives of the timber elephants over a number of studies. She found that infant survival (up to the age of five) was associated not only with presence of the mother, but also the maternal grandmother and siblings. I knew why Mirkka would be interested in quantifying that. The point underlying it all is whether relatives can in some way contribute to our doing well, and it makes sense in an elephant or a human, because we live in those groups and if siblings or grandparents could gain some additional fitness by supporting us, they could potentially buffer us from big changes in life. Mirkka found that being captured from the wild, like Thaung Sein Win, rather than born in captivity, like Hannah, increases mortality rates. Thaung Sein Win was captured when he was very young, which actually means he had the best outcome, increasing his risk of mortality the following year by just 2 per cent or so. The older an elephant was when captured, the higher this mortality bump was, reaching almost 6 per cent if the elephant was captured at age 40. And the capture effect persists over time. For Thaung Sein Win, the excess mortality would be under 0.1 per cent a year by the time he was 14 or so, whereas the effect would stretch out for over ten years for his mother, who was caught as an adult.

As humans, we are no strangers to the long-term consequences of our early life conditions, or huge events in our lives. These can be linked to health, just like the well-known correlation between birth weight and cardiovascular mortality in later life. The psychological impacts of life events and how they interact with health are really difficult to untangle, but also an ever-present facet of our lives. I wasn't thinking that much about the long-term health of timber elephants when I had my own ultrasound scan in Sheffield in 2015, after I finished my PhD and work in Myanmar. Perhaps I thought a little about the scans I had been taught to do on elephants in Thailand by Chatchote and Nikorn. How these scans were always quite exciting, looking for the potential of life in sex organs or a new one in utero. My scan was more mundane and the growth in my uterus most unwelcome. It was a polyp, just a benign thing I was getting checked out before it could be excised, and I could get on with what I really wanted to that evening, my leaving party. A year after completing my PhD and after assisting on Virpi's team for five years all together, I was striking out on my own. I had got what I'd always wanted. I was going back to Cambridge as a fellow.

I felt like a male elephant breaking away from the family and going out into the world.

My path there went right back to the first time I went to Cambridge as a 15-year-old. It was a cold December and there were no students around. Our footsteps echoed on the grey pavement. My dad had suggested a day out and I wanted to shop for a dress for a New Year's party. But there was something monumental about the place and there was great fudge, so just looking around was much more exciting than dress shopping. I saw a figure crossing the sprawling lawn at King's College, where the neoclassical Gibbs building sits beside the famed chapel, gliding past those signs that warn 'Do not walk on the grass' in several languages. This was someone who was above such mundane ambulatory instructions. They were practically walking on air, propelled by Darwin and Milton and Newton, skimming the surface of this world and deeply embedded in another scholarly one. This was the place where people debated philosophy over dinners with five different forks per person. Oh, I said, I wanted to study at Cambridge that day. But what I really had my eye on was a fellowship. I wanted to be able to sit on the high table with the people who formed a chain back to the founding of the university in 1208. I wanted to wear a long black gown and look like a bat as I wafted along. I remember looking at my dad, and I knew in another life, with a different configuration of opportunities, he would have loved to have gone to university. My grandparents would have too, but their families couldn't even afford the uniforms for them to finish grammar school. That day I promised myself that I was going to take them all to Cambridge.

And although some of the Cambridge magic rubbed off over time, it was still my dream to be a fellow. I still applied to every single college fellowship (at Oxford too) two years in a row and received rejection after rejection. With most, I fell at the first hurdle, for others I progressed to having my written work reviewed. For two colleges, I was invited to interview. I failed to impress at the first interview, having mounded all of my expectations and hope onto it. I remember receiving the notification that someone else, with a better CV and better publications, had beaten me to the start line. I sat in the same park where I'd eaten limpid sushi before meeting Virpi and wept inconsolably. But I picked myself up and applied with a reworked project and renewed enthusiasm the following year. And I got an interview at a different college. It was an extremely intense experience, meeting all of those fellows around a table. I had memorised a presentation, timed precisely to eight minutes, and recited it while pacing up and down my room tens of times the night before. Unusually for me, I did not vomit, but I breathed slowly and deliberately, reminding myself it was necessary for the 'keeping alive' portion of the interview. Afterwards, I stepped out of the room and tore away. I stood by the college war memorial as a crowd of visiting schoolchildren walked by and I tried to limit the amount of swearing to myself I was doing. I decided to cancel tea plans, buy a huge amount of chocolate and get the hell out of there. So I sat on a train just outside of Ely trying to

A family group in South Africa.

A young savannah elephant reacts to the field vehicle in South Africa.

Bulumko

Sizwe, note his broken tusk.

These young males came to the waterhole at the same time as Sizwe. The order in which they access water and who gets the shady spots by the water holes gives us information about their relationships.

Wild Spirit

We saw this family group at the end of a long field day in South Africa, cooling down and playing in the water.

Soshangane in 2015. He was upwind from us when this photograph was taken, and slightly nervous that he could hear us, but not smell us yet. (Copyright Moritz Muschick)

I took this picture from a hide. After waiting for hours watching everything from warthogs to giraffe come to drink, this male showed up and made my morning.

Elephants coat themselves in dust for protection from the sun and biting insects.

These two were associating with Intwandamela. The elephant in the foreground shows a distinct notch in his ear, something we would always photograph to help with identification.
(Copyright Christin Winter)

I took this picture in the Okavango Delta in Botswana. I watched the family group cross the river from the *mokoro* (boat) and then keep an eye on me before disappearing into the foliage.

Elephants can remove bark
from trees with their tusks
to eat, leaving scars like this.
(Copyright Moritz Muschick)

Dung beetle rolling elephant
dung in Africa.

stuff another chocolate éclair into my mouth when the phone rang. I was going to be a fellow! I could barely answer through my toffee-fused jaw and utter disbelief, but I somehow got across the message that I would like to accept. Now there was chocolate for everyone in the carriage to celebrate.

It was perfect. But I needed money for the elephant research. A few months later, I walked through a different town, Zürich. In a villa at the top of a hill, I presented ideas for my work. I got the second fellowship, the funding for up to five whole years. My luck had taken on a trajectory of its own and I felt like a passenger. After these successes, I took a holiday to the USA and looked past my toes and over the edge of the Grand Canyon. I could see the opportunities opening with the expanses below me. I could almost make out elephants traversing their way down the rock in a line. It was the most swept away I had ever been. So when I had that little scan for the polyp at the hospital in Sheffield a couple of months later, I didn't think much of it. When I was referred for a second scan because the first was inconclusive, I was faintly exasperated, but nonplussed. I went alone and chatted away as the doctor, this time, checked me out.

I barely noticed that she had stopped talking to me and that her expression had taken on a look of intense concentration. She was scanning higher up on my abdomen and pressing hard. She asked me to breath in and out. She wanted to check from the back, she wanted to look at my liver. By this time, I was not only late, but panicked. I imagined the news coming, I was pregnant, I had a massive worm, an alien, a dead twin in there. What was it? I remember what the doctor said despite barely remembering anything afterwards.

'You're going to google this and be scared,' she said. 'But has anyone told you that you have a kidney disease?'

People at the hospital talked to me about referrals and consultants and check-ups and said I must talk to my family because this is a genetic disease and they might have it too. All that I knew was there had been a mistake and I had to go to a party and then I was going to be a fellow. I left. I stood at the bus stop and the bus didn't come. I went to a shop and bought six bottles of water and drank them one after the other, thinking I might be able to flush this problem out if I drank enough. I arrived at my party, all of my friends there to wish me luck. I stood in the doorway and burst into tears.

One reason I think I am an elephant is because I have had those two lives, like Thaung Sein Win or the captured white elephants. I had the one before and the one after. Of course they are both my lives, but they barely feel connected. My first term as a fellow wasn't the dream I imagined it would be. Nothing ever is. But I didn't think I would have my first college meeting with a bruise forming under a piece of cotton wool taped to my arm from a slightly botched blood test. Or that my GP would have to convince me to put on my gown and go to the chapel to become a fellow, because staying in the surgery until someone told me exactly when this disease was going to kill me wasn't going to bear fruit. I don't want to be dramatic and present this as the worst news I could have had at the age of 29. The fact that I'm sitting here today and typing this is evidence of that. But it definitely changed me. Finding out I was the only one in my family with the disease was a huge relief, and selfishly, I noted it gave me more potential kidney donors in the future.

Suddenly all of those elephant lives, those deaths and births in a spreadsheet, were more than just data points to me; they were individual stories, played out thousands of times. I felt connected to them. And somehow, even though I lost the first three months after my diagnosis in a thick memory fog, I found a way to keep going. I got my dog, one of the best decisions I have ever made. And I got out of bed to feed her, and we walked the same route through Cambridge parks every day. The seasons changed around us, I saw herons and a kingfisher, but following that path was more a meditation than anything else. Our footsteps gave a rhythm and a shape to the days when I thought I had drifted from everything.

I had to think about the elephants I knew, like Bulumko who was blind and who still talks to me. And little baby Hannah who could grow up to be anything. They had their weaknesses, but I never thought of them as mistakes; Bulumko was more special because of his condition, not less. I thought of them as thick-skinned wonders. And I had to be like that too. I had to tell myself, you are not a mistake, you are a whole person. Your DNA typo, made before you were even born, isn't your defining feature. So I didn't just stay on the sofa. I did my fieldwork, wrote and won grants, moved countries and continents, published papers and taught students. I got through my first haemorrhage, which happened when I was attending an academic symposium in Zürich. I felt a sharp pain on my right side build over a day and then disappear, to my great relief. I went to the loo and produced urine the colour of red wine. I wanted to faint, but I couldn't. The hotel concierge gave me an aspirin, and I wished I was talking to someone I actually knew. A professor drove me to the hospital in his very nice black car with shiny leather seats that I really didn't want to get blood on. I reflected that at least it would wipe clean. The next day, I still gave my talk about elephants.

The elephants carry on too. Luckily, Baby Hannah is still alive, and well. I'm told she's rather bold; perhaps I should try to be more like her. I sometimes wonder how her life will play out. A timber elephant in training in a world with much less timber. Will Hannah drag logs one day? Will she have calves of her own, and be a more successful Hannah than I am? Hannah is an individual and her life might not play out along averages. She might struggle with disease, digestion, feeding her calves. She might live long, beyond me, or we might both grow older together, with our own scars. Just a year older than Hannah, Nyein Carly is having her life play out before us. On my office wall I have a watercolour painting of Nyein Carly as a one-year-old calf and her mother Aye Mya Mike. In copper tones, Nyein Carly strides out and her mother has her head and trunk turned towards her young calf. You can tell she is going to touch her. When I got the painting, I wrote to Carly to tell her about it, and she was touched. She told me that, at the age of six, Nyein Carly no longer has a mother. Aye Mya Mike died in 2017, of parasites.

CHAPTER 6

A team of rivals

Tangerine, Tangerrrrrriiiiiiiiiine, living reflection of a dream.
Led Zeppelin

I turned the Zeppelin up. Colorado was one of those places where you really feel there's space to crank up the classic rock and then lie back after you put it on repeat. It's where the plains hit the Rockies, and everyone's a mile high or more. I wasn't used to this kind of expanse, particularly when it's not punctuated with elephants. So, on the bus from Denver to Fort Collins, after laughing along when the driver asked me if I was from around here, I always put on my headphones with the volume up high to fill the space around me. Denver was shopping malls and shadows of beat poets to me, but as for Fort Collins, I didn't have anything to put on it. The craft beers and the twinkle lights on the high street, the cookie shop and the roadmap like a grid of tree names, and people playing basketball without their shirts on next to one of

Warhol's cans of soup – I had wandered onto the set of idealised small-town America and I was playing Confused English Person #1. I went to open a bank account, because I was going to live there for a while. Everyone in the branch was emphatically polite to me. I started to view them with an air of suspicion. What uniform were they wearing under that nondescript beige trouser and slightly less beige shirt combo? No one is that happy to help someone with $80 open a bank account. When the clerk told me I should take his card and call him any time, that he'd be happy to hear from me, because this relationship was important to him, I thought I'd not had that kind of commitment from men I'd been going out with for years in England. All I needed was $80, and to cycle one of those scary bikes (the ones you peddle backwards to brake) along some really wide roads where not even I could cause an accident (except when I cycle on the left), and I had my dream man. And he came complete with a drive-in ATM! I took photographs of that ATM; that's the real American Dream right there. I mean, I didn't have a car, but I felt as though I had made it.

There were a lot of wonderful things about living in Fort Collins. Like the surprise of finding one of my old friends from undergraduate days living there, and acquiring some new ones. I liked to travel from there too, when the experience verged from sweet to saccharine and I had to feel the space again. So I swam in the pool at a beautiful house overlooking Santa Barbara, among larger-than-life art and cats with over-sized personalities to compete with it. I wrote some of this text on the train from LA along the coast, a girl between the ocean and the tinder box, thinking of elephants. I drove from Fort Collins to Aspen with a friend one gorgeous crisp autumn day and we found some of the most beautiful orange-tinged bird feathers I've ever seen. We held them up as we watched the sunrise. I ate ice cream on the university steps on a Saturday night and watched trucks with Trump banners come in from the countryside and drive along the main street. In the ideal university town, with jazz in the square and chickens in my backyard, upheavals in the world were haemorrhaging into the picture. But even in an America at odds with itself, there was a question in the background. A prominent one, not the kind of unspoken, back of the mind one. One about me. It wasn't even the one I had to answer every day ('Are you really British?', of course not, I'm auditioning for a play), but: 'Why do you study elephants in Colorado?'

Colorado actually connected me to Samburu, the place where my life with elephants started. The professor I went to Colorado for, George Wittemyer studies the ecology, social lives and movements of animals, including elephants: those in Samburu and Tanzania and even in Myanmar. I was among elephant people in the heart of America. I had become more interested in the lives of male elephants through my studies and experiences. George and Shifra Goldenberg, who did her PhD there and was always a very measured and able researcher, had studied the social lives of Samburu males. In order even to begin to understand them, I had to strip away my precon-ceptions about males. I had always constructed them in opposition to females. Females had been the starting point for me because reproductive investment is easier to measure (you can count how many calves a female gives birth to, but there's more uncertainty in identifying paternity). It seemed that being a female was about kin bonds, raising offspring, and cooperative interactions with other females. In contrast, male life was about leaving the family group, sex and competitive interactions with other males. I was missing the point again: that male lives are lives in their own right and not only meaningful in how they diverge from those of females.

Take that phenomenon of musth, the surge of male hormones that becomes an annual occurrence in adult male elephants. To an extent, it punctuates male life, but when we read about the physiology and behaviour surrounding musth, we only think about when a male is in musth. Most of the time, he isn't. And there is a life being lived in those 'between times'; he's not just suspended between musth bouts, anticipating the next rush of hormones. What are the males doing then? How do they live and behave and eat, and where do they go? Shifra and George were interested in how males associated when they weren't sexually active and their associates weren't either. During these times their behaviour was quite different. Male African elephants are probably much more social than we previously thought, because we just weren't taking into account the association index (a measure of the frequency of associations). If we think of an elephant social network, if we compare the sexually inactive (not in musth) to the sexually active (in musth) time, males had larger networks of associates, and also denser networks. Males also preferred companions closer to them in age when they were sexually inactive. What I found really striking was that, when sexual activity wasn't considered, there was no difference in the number of associates (other elephants they hang out with) a male had depending on its age. When sexual activity was taken into account, both in terms of the elephant himself being in musth and the other elephants too, the number of associates increased with age in adult males. I think we often worry a lot about 'overfitting' in science, that we throw too many explanatory variables into a model and lose our ability to explain any variation. But this seemed like a clear case of something that needed to be in the formula because it shifts everything for a male. This is definitely not the final word on male social behaviour, but it does highlight that we can't just transfer across what females do and repeat the analysis on males, because we might miss some critical patterns.

I also thought a lot about how those social bonds that males forge might be reflected in kin bonds. In female elephants, they align closely, with close relatives, such as mothers and their adult daughters spending a lot of time together. More distant associates are not so closely genetically related either. This is all super straightforward for us to explain. Relatives associating, protecting and helping each other, helps them look after the babies in the group. So it means that their shared genes progress to the next generations. Easy. But what about males? Do they hang out with relatives after they leave their family herd, in their adult lives? Fathers aren't involved in parental care in elephants, but could there be a possibility of young males staying with relatives of their own age, or finding older males they're related to when they disperse? I came across another study from the Amboseli site in Kenya. Patrick Chiyo and his colleagues there found that there was a link. Although not as strong as in females, they did observe a positive relationship between genetic relatedness and associations (hanging out together), with genetically related individuals having a higher association index. Like Shifra and George, they also found males associated with age mates. And that males of all ages do associate with other males. They explained this was potentially because males sparred with others close in age to them; so essentially testing out their fighting skills in a safe social context. It sounds a bit rough and tumble, which is completely the case. If you watch it though, it can be beautiful; they can pick up the dust as they dip down their head and make a starburst as they lift their head back up and shake it out, ears out, looking even bigger than you thought they could. And that's just the display version, the gentle version. You know you don't want to be in the way when he's not just showing you who he is. Perhaps the sparring is like humans doing a sport together, or even just play-fighting. We know elephants play and it's part of their development, giving them space to test out behaviour they might need as an adult. Chiyo suggested that sparring might help them to prepare for competition with males when they are sexually active. Such events represent much higher stakes: males risk their lives in those interactions, blood spilt by tusks can run in rivulets on the dry ground.

It all made me think of a male that I had met called Soshangane. After years of working in Asia and with an urgent sense of my own mortality, in 2015 I decided to go to South Africa. We flew into Johannesburg and whiled away a few hours at the airport, because, as usual, I had been over-cautious in booking connecting flights. And for some reason, I really like to build up the anxiety for getting on a smaller plane with pacing, caffeine and exhaustion. I'm not going to pretend I enjoyed the one-hour hop to Hoedspruit – that would be a lie. But the plane wasn't too tiny, or obnoxiously late, and there was the view. It was breathtaking. A long plain transitioned into waves of striated mountains. It seemed so vast and empty, but I noticed the road traced across it. Faint human lines in the dust. I hadn't been sure about going to South Africa. I didn't know anything about the place. It didn't seem as wild as Samburu; in fact, it seemed too real. Like many wildlife ecologists, I indulgently bask in my own vision of an imagined wild Africa that is frankly a disservice to the people and animals living there and to the complexity of the place. Being an ecologist is about seeing the wounds and the scars in a landscape, as Aldo Leopold said, and often that is incredibly lonely. But almost no landscape is untouched by humans, and not everything humans do has to come with a negative value judgement from me, nor does it have to have me bleating on about how some people are 'closer to nature' – whatever that would mean, coming from someone who flew in on a jet from London.

What South Africa, Soshangane and that aeroplane taught me was that history written in a landscape can be beautiful and challenging and painful, but I can see it is still a gift. History is also written on the faces of everyone you speak to because they're part of it, and the land without the people doesn't make any sense. It's fitting that I saw the area first from the air because whenever someone asks why I go to South Africa, I always mention the sky. The horizon is bigger there. Bigger even than in Colorado because giants still live there. This makes a place magical and mythological and sometimes bloody scary. When you're flying over the plateau, you feel as though it could go on forever, as if there's never been this much opportunity and potential. It is ecstatically and rapturously thrilling. You hit the mountains and the strata make you feel like the newest and least significant thing that could touch them. As if you're looking at another planet and it's all far beyond you. And then you start dropping in altitude and gripping the armrest of your seat because you have the lowveld stretching below you. It's savannah, but bushy, mostly at a low level, but dotted with trees, some with twiggy vulture nests up top in the canopy. You have to work to see an elephant there.

That's how I gave myself to that place, knowing my insignificance, my ignorance and my ultimate death are dwarfed by its bounteous horizon, and anything I did there would be huge for me and tiny for it. On the plane, that first time, I told myself, you just have to turn it up, Hannah: do listen to the news, do make friends, do not assume all your friends will be people who have PhDs and work in a job like yours, do let them talk to you about how it was growing up there, do let them show you their neighbourhood, do go to the braai (never call it a barbecue, even if you might know it as that), and to that dance. You should sit on the back of the truck and look at the stars through your binoculars because you will never see diamonds like that if you robbed all the jewellery shops and safes in the world. In the day, you should get up early, 5 am in summer, and wash, pack your camera, sunglasses, water and GPS and go out with the guys to listen for elephants. Maybe you'll see twenty, maybe just one, maybe only an ear or the bleeps from a tracking collar. Maybe there will be a hyena standing by the side of the road, mottled hair raised up, ear twitching, dry blood caked on her muzzle and the face of a lion cub in her mouth. But above all, watch and listen.

I was lucky that after such a short drive on one of those days, I saw Soshangane for the first time. He was a leader. Some elephants are. You just know. He was often in a gang of other males, which, like him, were in their twenties, perhaps six or seven of them at times, travelling together, or all drinking from a waterhole. He had stunning tusks. They thrusted forward, perhaps not as elegant as the low-sweeping classic Kruger style. He was handsome. Standing tall he had it all ahead of him. We were lucky because he hung out so close to the field base that sometimes we could tune into his collar frequency and find him within 30 minutes. One day we tracked him to a waterhole and watched him drink some of the 200 litres he might consume in a day. I tipped back my hat and took off my sunglasses and smiled. I know I shouldn't say it, but he was my favourite, ranging so close to where I slept. We were upwind of him, so he couldn't catch our scent to recognise us. Despite having that powerful sense of smell, even better than that of a dog, he couldn't place us. I watched him hold his trunk out and explore the air, then wrap it around and touch his ear, comforting himself. Hi Soshangane, it's Hannah, I thought. Could you please leave us a faecal sample? I wanted to test how closely related he was to those males he associated with, and I had a brand-new student called Tess to work on it with me. Shosh didn't disappoint. Before he left the waterhole, he defecated. While I waited for him to disappear into the horizon, I snapped on my gloves. We were not in Myanmar at a logging camp; I had dung beetles to compete with and no oozies to whisper in the ear of the elephants and keep them calm. As I bent down, I saw one of those dung beetles, like a dirty garnet on the side of a bolus. A jewel in my gold. I watched its back legs working away, shaping a smaller ball from the big one. Then I got my sample from another ball. There was enough for everyone.

I was being introduced to the elephants by a non-profit again, this time Elephants Alive, which was linked to Save the Elephants, the people who hosted me in Samburu. The number of elephants in this corner of South Africa was healthy, where the private reserves had dropped their fences to the mighty Kruger National Park. Males had moved into the area first, but now it buzzed with herds and younger males like Shosh. And it wasn't just elephants. I saw leopards, wild dogs, hyena, white rhino, hippos and giraffe there. It was different from Samburu: the giraffe were paler than the reticulated giraffes I had known in Kenya, the ostriches didn't have blue legs, and there was no cause to shout out 'dik dik!', because their range just didn't extend into the area. As for the rhino, they had disappeared from Samburu, but I saw many of them in the Kruger area. I took my parents out there, and my mother even spotted two black rhino when we were on a game drive there. She was perhaps not as excited as I was that there was an excellent spotter in the family. They're just like rocks, but some of them are different, she explained, shrugging. My dad lay down in the back, enduring a few bumps; there was no way he could compete with that. I thought that there was something incredible about the density and variety of wildlife being so close to the town, Hoedspruit, which had fast-food joints and a supermarket that felt so stark. You really felt the contrasts there.

One day we were driving along the main tar road. It was winter, cold in the wind and about 6.30 am, perhaps a little later. We saw a couple of vehicles pulled up ahead. Not unusual. This was a safari area. We would always slow down and make sure they had the best possible sighting with minimal impact from us. We would stop for elephants out of the way and collect all the data needed – ID photographs, number of elephants, what they were doing, where they were going – and I would always listen out for that thud, thud, thud, the dropping of elephant dung. But this wasn't an elephant they had stopped for. We were closer now and there were no grey giants hiding between the trees. This was something low. The people by the vehicles waved us down, which was unusual, telling us to stop too. And then I picked up the word they were mouthing to us. It couldn't be, I thought, there's no way. Right by the road! It was a pangolin. I knew I worked with an iconic species and a trafficked species. I knew that there were bits of elephant all over the world, playing music, celebrating god, stamping names, being beautiful, being many things other than part of an elephant. The elephants they originated from were reduced to outsized shadows of death behind all of the lovely things. Things that were so white, but you're sure you sometimes see the blood dripping down them.

Pangolins, like elephants, are also trafficked. They are also in demand for their parts. I understand why they are prized, these quirky scaled mammals, but disappointed that we kill them for it. When I looked at that pangolin under the bush I knew it was otherworldly and that nothing I had ever done could render me deserving of seeing it. With the scales covering its back, it was like a pine cone made into a suit of armour. But it was alive and a part of nature. Its little black eye caught mine. I felt bad for the pangolin. Unlike elephants, pangolins are not huge enough to terrify people out of following them.

They are rare, but easier to snare, to trap, to kill. Their defence mechanism of forming a ball is very useful for some predators, but just makes them immobile and easy for humans to pick up and carry away. They also lack the grandeur of an elephant, the instant recognition. And so perhaps they have less attention, less light shone on them, and the death shadows they cast across the world are shorter and not so prominent. But on seeing that pangolin, I knew the human in me wouldn't let me forget it. And the elephant in me saw across the divide.

In Samburu, we had driven along specific routes we call transects to find the elephants. Here it was more complicated; there were more roads, more elephants moving in and out and plenty more bushy vegetation. So we often used the collars to guide us to elephants. They gave a signal every hour with a GPS point, which updated on an iPad that the team carried. The collars also had a VHF signal so that we could track the old-fashioned way: by holding out the antenna and picking up the signal, we could hear a bleep if an elephant was within 4 kilometres, and the signal was stronger the closer we got. We weren't always that dependent on the collars, though; sometimes eyes were enough. I remember seeing a slightly battle-hardened and craggy male named Wild Spirit when I was out with a field assistant. He had a collar on, but the signal wasn't being emitted. Fortunately, she had sharp eyes and spotted Wild Spirit in the shade of a tree. Like Shosh, he was spectacular. But in a different way because he was a little older and more experienced. His torn ears hanging like well-thumbed book pages.

I saw the hole in his left ear and a distinctive tear right next to it. These elephants did not come with logbooks and ID numbers to tell you all about their history. You had to look for it and read it in them. Wild Spirit's tusks were pointed at the end and thick at the base. His face was wide, a reflection of his age. But all elephants are different. Take Intwandamela and Sweetie, two others we were following in the area. Intwandamela had a thick and heavy body and short legs. On the other hand, Sweetie's tusks never grew long, which gave him the appearance of an elderly individual who needed false teeth. If you watched him, though, he could move a younger male away from the freshest water, or away from the most refreshing shade. Just because he didn't have the biggest tusks didn't mean he lacked dominance. I tried to recognise all the elephants based on their appearance, but was nowhere near as good as the team who had followed them for up to 20 years. Their commitment to the place and their focus on detail was incredible.

One day we had a pleasant surprise encounter – no collars needed. We found Sweetie and another big male, Napoleon, with extra-thick tusks, by a swimming pool. Not a river or waterhole, but a human-made pool for humans. The water was azure. The comfortable chairs and sun loungers were empty. Sweetie sucked up water with his trunk and quickly blasted it into his mouth. Behind him and Napoleon, among the ones we could easily identify was a bunch of younger males. They ranged in age from late teens into their early thirties. At one point, one male placed the vulnerable tip of his trunk right by Napoleon's mouth. They all moved slowly and easily. The group shattered my ideas about antagonistic and aggressive males, just like Jack (from the Okavango Delta) or Soshangane and the males he hung out with. I took a photograph and wrote on the back of it: 'Boys by the pool, drinking, October 2015.' Because that's what it was. They were calm, relaxed, totally unlike me. I was nervous to be just 50 metres or so from them and on foot, and it was the first time I'd been around so many adult males. But they weren't skittish at all. They drank, touched and stood for an amount of time I could never quantify. It felt like decades, but was probably about 20 minutes. One elephant rumbled just once before slowly leaving, meandering away. They were less vocal than the females, but they were still very much a group. After they had left, I walked up to the pool and peered at my reflection. The surface of the pool was as calm as they had been.

Even though not every elephant had one, the collars were incredibly valuable. One thing about them was that the data on location updated all the time, so you didn't have to have a researcher with the elephants to know where they were. The patterns of movement with the season, with musth and with resource availability become apparent in the form of coloured lines on a map. Even if an animal went into the neighbouring Kruger National Park, or in those problematic cases when it left the protected area, it could be tracked. So the collars were definitely worth the investment from both an academic and a practical perspective. I was fortunate to go on the incredibly well-organised and professionally performed collaring operations with the Elephants Alive team. I couldn't sleep the night before. The very idea that the team was going to dart an elephant from a helicopter gave me a rush of adrenaline, even though the pilot and veterinarian were extremely experienced, even though the elephant had been tracked by vehicle for days, even though everyone knew what they were doing and were relaxed. It wasn't just exciting; it was more than that.

On the ground, we drove to the area where the elephants were being tracked. It was South African winter and the boys sitting out on the back of the *bakkie* were bundled up in scarves and hats and still looked freezing. We had a debrief and safety talk with all the participants. The veterinarian patiently explained the darting and the collaring procedure. How we should stretch out the trunk of the elephant, how we would keep the airway open, how we should quickly perform all our tasks and he would administer the antidote, so the elephant would get back up. We were collaring three elephants today. Dr Michelle Henley, the Director of Elephants Alive, told me it would be done by lunchtime, and I was impressed. Things went along smoothly; the beating of the helicopter above me wasn't agitating, but a comfort as we bounced along the roads, the radio crackling, and we received the message that the first elephant, who was yet to be named, was down. As we pulled up, the area was already safe and the helicopter team were with the elephant, making sure it was resting safely on its side. They had stretched out his trunk and kept the airway open with two little sticks. So far, so good. We worked diligently around him, about ten people altogether, ants in an elephant-shaped colony. I helped take measurements of its supine body and took the liberty of getting a faecal sample directly from its rectum, dropping it carefully into a tube filled with salt solution, while the team also took body temperature and blood samples. Most importantly, the collar was fitted carefully in place: it was not tight or heavy in elephant terms and would not restrict movement, slow an elephant down or affect feeding. Sparks flew as the bolts were fixed to hold it in place and the excess strap was cut off to prevent any irritation. The counterweight would ensure that the signals would always be directed upwards, beyond helicopters and aeroplanes, to a satellite in the sky. Elephants in space.

The whole operation had been so slick that it was only when I pulled off my gloves and the elephant was still down, that I realised we were with a sleeping giant. Closer than I had ever been to a live African elephant. I looked at the bottom of his feet and toes, and placed my palm on his foot pad, which was both smooth and ridged at the same time. I put my hand on his chest and felt it rise and fall in that slow-motion elephant pace. His papery ear had dropped a little over his face. He smelled like an elephant, he still was an elephant, but not his full self. The veterinarian administered the antidote, and we ducked back behind the doors of our vehicles in anticipation of him waking up. The helicopter would fly off for the next collaring, and this male would contribute data for the next five years or more. We would get back in the vehicle, and he would pull himself up again, hold out his tail, slightly confused, and look around towards the dry riverbeds and the other elephants and the high sun. That's when he'd be a full elephant again, when he walked away from me. Now his name was Dumo.

It's strange to think that Dumo will be sending signals as he walks now, wherever he goes, right out across that bushy landscape. And perhaps not just across the open sides of the reserves, where they lead into the Kruger National Park. What if he walks the other way? What about the roads, the fences, the farms? Like a lot of people, I thought there couldn't really be issues of humans and elephants competing for the same land and the same resources when there was a fence between them. The fence was the solution. We're not talking about a line in the sand – us on one side, you on the other. It's a structure, it's physical, it's a barrier. Fences seem solid; electric fences seem impassable; you are safe behind one, and what's on the other side is safe from you. But of course, it isn't that way. An elephant can walk with the tip of her trunk along the fence, not touching, but close to the wire. She knows if it's live. An elephant has tusks that don't conduct electricity and the electric wires can be pinged and snapped. If an elephant is really intent on going somewhere, it will find a route. Same with humans; they can move in the opposite direction with wire clippers, a spade, a snare. Either side of the fence, a breach might incite fear, disruption; lives are impacted, there is always a fallout. This is not a hard border. It's a porous transition zone. This is where it is difficult. It is the lightning rod for the challenges of humans and wildlife living in the same area.

One day I was sitting in a hide waiting to see if elephants would come. None had so far, just a few impala, and a warthog family – it was the time of year that they had piglets. I always associated Hoedspruit with warthogs, because I saw them foraging by the roads, in the wildlife park where people lived. I even bought a metal warthog from a roadside vendor. He wrapped it in newspaper, but I still cut my finger when I unwrapped it, dripping blood onto a report about student protests in the city, Johannesburg, far away from here. That morning, the piglets made me grin, with their tails up in the air like antennae. The hide overlooked a shallow waterhole, and the birds sang above us. I waited for several hours, watching the light shift. I tried to identify the trees with their leaf shape, and their bark. I gave up after getting the easy ones and wrote *haiku* in my notebook. About the crane that did not move, pinned in the world that shifted around it; about the smell of the rain that is really the smell of the ground that you only notice when it gets wet a little way off, and you miss it so much, you want the rain to come to you. And then when the rain was arriving, turning your face to it and letting it splash all over it. I was perfecting a series of *haiku* about the dead frog in the corner of the hide while I curled and uncurled my toes. Sometimes it was good to watch and wait. But today we couldn't do that all day. Michelle's car came pacing along the road, kicking up a small plume of dust behind it. She stuck her head out of the door. 'You have to come now,' she said. 'It's Derek.'

I have a lovely Irish post-doctorate researcher who worked for me called Derek Murphy. He has a massive amount of energy and runs most days, even when we're in South Africa. He also has impressive technical skills and he uses supercomputers to run really complex analyses of elephants' social lives. He has included musth in our analyses of male elephants' bonds over time, showing older males have more consistency. He is a very nice human. But this was about another Derek, an elephant. Elephant Derek had form. A year earlier, I'd been in South Africa when Derek was causing problems. He was routinely getting out of the fenced reserve area, scrambling down a steep riverbank and walking along the river to a mango farm. There he ate mangoes. It sounds cute, but it was serious. Oftentimes a handful of other elephants went along with Derek. Michelle had to walk a tightrope of keeping the elephants safe and also listening to and acknowledging the concerns of the farmers. She and the field team, including Jess, the relaxed but always prepared South African researcher who took me out many times to see Bulumko and the other elephants, hatched a plan. They used chilli, a well-documented elephant deterrent, to keep Derek away. For several days, dry elephant dung balls were injected with hot, dry chilli by the Elephants Alive team, and at night Jess slept nearby, watching for Derek and lighting the balls, then keeping them smouldering. A pungent smoke hazed the area. No animal with as sensitive a sense of smell as an elephant could tolerate it. It worked. Derek did not go to the mango farm. But after a couple of days, Michelle got a disappointing call, saying Derek and his crew hadn't returned to the reserve. Instead, they had started to frequent the butternut squash farm.

To be clear here, there was enough food in the reserve, but at the end of a dry winter, perhaps the bark, acacia and sprinkle of leaves and grass weren't so appealing to Derek. Once he got into a behavioural pattern of going out beyond the fence, what was going to stop him? The food was abundant, cheap (he didn't have to spend a lot of energy getting it), tasty, calorie-dense. And Derek was young, in his early twenties, still defining his home range, musth cycle and patterns of association. He was still figuring out how he was going to be an adult male elephant. Perhaps this was a short cut to get him big and bulky enough to compete with other males. Researchers in the Amboseli area had shown male elephants that ate crops there had larger body sizes for their age, potentially giving them an advantage in competitive interactions with other males. Michelle made a plan to collar Derek and his two buddies, then she would know when they approached the fence so they could be moved away. They would also put sharp stones along the area where he usually got out, so it was no longer an option. It seemed harsh on Derek, but he had to learn this way. He was putting his life in danger by walking in areas with roads, and he was also a risk not just to livelihoods but to the lives of humans who might come across his path and spook him. It was a very complex collaring operation, as the three elephants had to be loaded onto trucks and moved back into the reserve when they were collared. But Michelle and the team pulled it off, and for a year, Derek and his elephant avatar on the iPad had stayed safely within the confines of the reserve.

Today, though, he was back out. Michelle was already deep in yet another bout of decision-making. We needed to cut the grass and clear an area so that a helicopter could land at the field site. A car would be sent out to look for Derek so we had eyes on him, while Michelle contacted the various people who needed to be aware of Derek's whereabouts. I went with the team to look for Derek on the ground: we followed his tracks, listened to the word about him, but ultimately could only narrow his whereabouts down to a wide area. We put it off until the next day. That day Michelle was up in the helicopter. They would herd Derek back into the reserve, along the river, the way he had come. I sat on an old railway bridge, one that I had walked along very tentatively, seeing the river below between the rails and holding onto the metal girders. I imagined looking down from a train. The ground would clip past, you wouldn't notice the young male elephant in the riverbed, and perhaps you wouldn't know that the river was named Oliphants after his kind. Soon you'd be over the wide bed and going up north, maybe to the mines, perhaps to Zimbabwe, maybe deeper into Africa. I noticed every step, looking through the gaps between the tracks and seeing the grass and then the sluggish water, aware of the drop and the mess I would make if I fell through. But we were here for Derek, not for me to make an exhibition of myself.

He broke my train of thought. I heard the beating of the helicopter propeller, and, below me, there was Derek moving quickly along the river, ears splayed and tail aloft, coming straight towards us. He appeared a little agitated, sticking close to the cover provided by trees. He was nervous of the bridge, even more nervous than I was. He was alert. He didn't like the people 30 metres or so above him on the bridge. We got down, got out of his path. This was the only way to get him back into the reserve. The helicopter had a flashing light and a siren, like that on an ambulance. I saw Michelle leaning out of the helicopter, making sure they kept track of Derek, making sure he was safe. Eventually, he was returned to the reserve. The next morning, the team got to work putting more rocks in place to ensure that this was no longer an attractive exit point for Derek. I didn't want to leave. It was incredible, I told the team. He's back now and it could have been a disaster. Everyone just smiled at me. Just another everyday miracle.

On my first trip to South Africa, Soshangane made me think everything could work. He was easy to find, he always had his crew, he rumbled, making me believe I could make my observations and my sound recordings, and collect all the dung to construct his social and genetic networks. Soshangane sold it all to me months after I'd put funding proposals together and sold the idea to other people. On the second trip, I was so excited to see him. I had all my equipment now. It was all systems go. Except it wasn't the way I wanted it to be. In his prime, Shosh had died. It's not another grim poaching story, it's just part of elephant life. Shosh had been in a fight, a big one, and he'd got a puncture wound from the tusk of another male. While Shosh had shown me the affiliative, collaborative side of male elephant life, I couldn't fool myself into thinking there was no competitive side. It's a disservice to gloss over that. Some element of between-individual physical violence is part of being a male elephant. And perhaps only a small proportion get into fights the way Shosh did, but we know that it's part of elephant life even without humans in the picture. They're not some ideal we can hold ourselves up to and imagine how good they are. They're individuals making conscious and unconscious decisions. They struggle. And Soshangane had fought. His wound had probably become infected. He'd died, young, doing what a young male does; he died the elephant he was. But I couldn't think any of that when I first heard about it. I just felt my eyes fill and swim and then overflow with tears. How could he leave me like this? All the things we were supposed to do and see; I was going to document his life and then he goes and ends it because he's a male and that's what they do. It's not all boys drinking by the pool. I was angry with him. He had been so beautiful with his fine tusks for one so young. His imposing size marked him as a future big tusker. He was going to have some of the most impressive tusks in the area if he lived longer, tusks people would look at and shake their heads in disbelief. Except he wasn't now, because he died before he could ever break one clean off in a fight, or before his eyes became cloudy with age and his face wide like a tree trunk. I picked up my equipment in the morning at the Elephants Alive office and walked past Shosh's bright white skull outside. Boy, you're gone, but I'm going to do this anyway. You sit there and watch.

By the time I went to Colorado we had, for many months, patiently waited for elephants to poo, watched them hang out by the river, by the water, on their own. I had forgiven Shosh for being the elephant he was and celebrated the fact that he hadn't died at the hands of a human. We had made our recordings and had mostly been successful, and we had also made recordings ruined by the wind, by engines, by a plane landing, by me talking over it, and by a really loud bird. The data started to take shape on (human) Derek's computer screen, and my student Tess worked away in the laboratory, weaving scientific gold out of elephant dung. Derek showed that there was a lot of fission and fusion over time in groups of males, like the ones we so often saw around Soshangane, but also that we have to be really careful in comparing populations. In Samburu, we had driven set transects through the landscape; in South Africa, we tracked elephants using their collars. This meant we got different perspectives on the elephants and their social lives, the first way giving a broader view, the second focusing closely on certain individuals. And Tess showed that, as we suspected, the elephants from the private reserves were part of a single genetic population with neighbouring Kruger. The spaces might be divided by humans in terms of management and activities, but the elephants were using it as a single landscape. That's why I was in Colorado, to talk about all of this and to thank the people who had started me on that track, inadvertently, almost a decade before, when I saw Yeager by the breakfast table. I loved the company and the excitement of people doing studies incorporating technology into conservation.

But one of the most striking moments to me came, as many do to an elephant person, in bullets. I woke up one morning, in my lovely little birdhouse overlooking the chickens, with the aqua-coloured big fridge and the comfy chairs. And I read that early that morning, four people had been shot close to my university campus. Three were dead. One of the dead was a student of the university. It had happened a kilometre or so from my little house. Far enough that I hadn't heard shots. Close enough that I felt it. I hung my head over my breakfast cereal. I knew there was real death, real shootings with elephants – I had seen it – but this was with people. With the elephants I always thought it was beyond me, and that there was nothing I could possibly do about it. But here I was, hanging my head again. I realised then that I felt most like an elephant when I felt loss.

CHAPTER 7

Elephant in the mirror

For I do not exist: there exist but the thousands of mirrors that reflect me.

Vladimir Nabokov

I travelled to northern Thailand in the wet season of 2011. I felt as though I had barely stepped out of the car when my skin exploded with mosquito bites. They were pink and hard like balls of bubblegum; my tight skin stretched over swollen lumps. My shirt stuck to me with a sweat that didn't dry. But I didn't care. I was meeting Josh Plotnik. Josh was researching elephant cognition there. He was a New Yorker and spoke as though he was impatient, as though he didn't have time to beat around the bush, as though he was going to change the world. He laughed at me a lot when I ordered hot black tea instead of iced coffee. I'd read his work, and expected a kind of professorial type, but I got a friend, a confidante, an innovator.

What surprised me was that, back then, he was willing to

listen to me at all. I remember walking around the village with him, close to his study site. It looked out onto the Mekong River, which at this point was wide and open like the best horizons. We could see Myanmar and Laos on the other side. I followed Josh along the pavement, weaving between the dogs, the plastic stools outside of restaurants, the adverts for boat tours and the parked motor scooters. Josh would greet people in Thai, know what the best food was to order and would balance delicately on a little plastic stool to eat it. The same stool I'd tripped over minutes before. I told him about my vision: that elephant life history had evolved in parallel to that of humans. But it was incredible that we humans had so much in common with the elephants in spite of that distant relatedness: the long life, the offspring dependence, the learning in a long 'childhood' phase, the high investment in calves. And we didn't share it because of a recent common ancestor. It was deeper than that, it was a shared strategy, a shared approach to a certain kind of life, and a rare one. He grinned at me. He'd been on a similar track, but as a psychologist. He was thinking of the convergence in terms of cognition and behaviour. From two disciplines, we had come to a similar conclusion. Potentially elephants make strategies like humans do, and it's interesting precisely because they're genetically distant from us. And we had found each other, on the banks of the Mekong.

As well as Josh, I met the humble and self-effacing John Roberts, who directed the elephant camp and the associated charity at the time and now works on conservation projects across the world. John's Englishness was a contrast to Josh's New York attitude. But John wasn't bumbling; he was sharp, even though he wouldn't admit to it, and he was very kind and understanding of researchers. He also knew the best local beers. He'd listen patiently to me talk about weighing elephants, balancing the pros and cons of different life history strategies, and he'd look at all the plots I made – saddle shapes, colour grades and error bars. This kind of data in the visual realm, the human and tangible kind, was my favourite, but not necessarily what everyone wants to be bombarded with. John took it all in. Josh and John took the time to listen to me before I had lots of letters after my name and before I had published any empirical work to back it up. They also expanded my vision. And they introduced me to some extraordinary elephants. One was Pepsi.

When I first saw Pepsi, he was a young male, on the cusp of adolescence. He was about eleven years old. He was already taller than my head at his shoulder. He was playful and very closely bonded to his mahout and the mahout's family. He used to wrap his trunk around his mahout's daughter and lift her up, making her squeal with laughter. He was also very good at picking up the discarded flip-flops of tourists, passing them gently to his mahout. I have a photo of him and me from that time. I'm grinning and throwing a peace sign, and there's a blast of wet-season green foliage behind us. I'm resting my hand on the top of Pepsi's head, feeling his fine spiky hairs. They're the texture of toothbrush bristles, but black and densely distributed. He has his trunk curled up in the air, his two tusks already prominently jutting out, his eye reflecting in the light of the camera. We're standing in a grassy area, slightly claggy with mud at that time of year. This is where Josh did his experiments. One of the first I ever saw was the 'elephant in the mirror' test. It's basically a test of self-recognition. Imagine you look in the mirror, you have a mark on your face, not a mole or something that's usually there, but something unexpected, say a painted cross. What's the first thing you do? Many people would touch the place on their face where they thought the mark was, given what they'd observed in the mirror. We understand that the reflected image is of us. It's not a different human, so if we see the mark in the mirror, it's probably on our face and we want to rub it off. And Josh told me that humans, being primates, like to groom themselves and are adapted to the visual domain. We rely on visual cues a lot. Getting rid of a mark using a mirror is a great self-recognition test for us.

Chimpanzees had also undergone this test. Josh explained to me that this was before research ethics, and the use of animals was not quite so thoroughly considered as it is today. The chimpanzees were actually anaesthetised before they were marked; this was so they wouldn't just touch the mark because they felt it being painted on. The researchers wanted to ensure that it was only in the mirror that the chimps would become aware of the mark. Josh wasn't able to do this with the elephants. Instead, he put a white face-paint mark on one side of the elephant's head and also a 'sham' mark in a glow-in-the-dark non-toxic paint on the other. The point was that the elephants could feel both marks being applied, but the 'sham' mark would be invisible under normal light. In different trials, he switched the sides to control for the fact that elephants might just tend to touch one side of their head more than the other. And that's not a totally ridiculous assumption, given that elephants can have a 'lateral preference'. Essentially, just as we're right- or left-handed, elephants are right or left 'tusked' and 'trunked', with a preference for one tusk for tasks like stripping bark from trees, and one preference for turning and gripping with their trunk over the other. This is often clearer to see in Asian elephants, because they only have one 'finger-like' projection at the end of their trunks, while African elephants have two. Asian elephants tend to press the projection down against the end of their trunk and then curl their trunk around to grip. They can either very delicately pick up items – not breaking a poppadum, for example – or they can show strength with the same appendage, such as pulling up a plant along with the roots. In contrast, African elephants can use the two fingers to pinch, as well as pressing them against the trunk itself. I roll my eyes when I hear about the uniqueness of opposable thumbs as a requirement for dexterity and fine tasks. It's obviously not true. The trunk is a highly sophisticated bit of apparatus and certainly an alternative to our hands. Interestingly, while the majority of humans prefer their right hand, it's not the case for elephants. It's much more like half and half, and elephants don't seem

to follow their mother's preference, that of their social associates or individuals they feed near. The pattern of variation
doesn't fall along age or sex lines. We don't really know what
determines the preference; we just know it takes place early
and then doesn't change. It's similar to the way that we often
struggle to write with our non-preferred hand if our preferred
one is out of action.

But back to those mirrors. Josh knew that the test had
been tried on elephants before, but the mirror was small and
out of tusk's reach. He knew he had to go for a much bigger
mirror, so elephants could explore and interact with the
reflection. When I arrived in Thailand, I saw one mounted
2.5 metres high. It wasn't glass, for safety, but a plastic that
wasn't warped. I didn't feel as though I was looking in a
fairground mirror when I looked at myself in it. I touched
the surface, and I touched my own face, sticky with sweat,
so appealing to mosquitoes, but perhaps not the best I've
ever seen myself. I have so many mirrors in my life that
perhaps I wouldn't be a good participant in a mirror recognition test anymore, I often want to avoid them, to shy
away. But I was here to see Pepsi, not myself. By 2011, Josh
had done the experiment plenty of times, with Pepsi and
with other elephants in Thailand and in zoos in the USA.
I'd read the paper, I'd seen the videos. Now I saw it for real
(and, of course, through my video camera). Pepsi had a
prominent white cross daubed on his forehead, and I couldn't
see the sham mark, but knew it was there. Pepsi sauntered
over to the mirror. I was expecting him to go straight for

the mark, but he opened his mouth. His rows of molars, that elephant conveyor belt rippling in his gums, were exposed – one of the five sets that would grow through in his lifetime. He pressed against the mirror, trunk aloft, forming with the reflection a kaleidoscope of pink tongue, off-white teeth and grey skin. I looked at Josh. He was nonplussed. This is self-directed behaviour, he told me. Animals often use the mirror to explore, looking in their mouths, their ears, splaying their legs and taking a look at their genitals. Following the brief but thorough dental exam-ination, Pepsi stood directly in front of the mirror, and, with the tip of his trunk, smeared the white cross on his forehead slightly. He passed the test with flying colours.

That night when I was excitedly discussing it with Josh, he was already on to bigger things. The mirror was just the start. You see, it showed the elephants might be aware of themselves, like other species that had 'passed' various versions of the test: dolphins, apes, killer whales, Eurasian magpies and even the tiny cleaner wrasse. However, it still has limitations. The fact that humans and other primates tend to like to groom, whereas elephants spend their time throwing dirt and dust across their faces was one; another was that we are so visual whereas other species might not be. They might be much more interested in olfactory, auditory or touch-based cues. I thought of the whales or the fish taking the test. They didn't have hands or a trunk to touch themselves, so the scientists could only measure how long they spent by the mirror, interacting with it, and whether this was different when they were marked or unmarked. It seemed to me like another instance where we were barely scratching the surface and limited by our own human-centric view of the world. And for Josh, no matter how many times we made him talk to us about that mirror, it was just the start of what elephants are capable of.

The next day, I went to see another experiment he had worked on. This was focused on cooperation. He'd had the idea after learning about experiments done by Meredith Crawford beginning in 1937 with chimpanzees, in the days before we carefully considered the ethical implications of animals in research. Chimpanzees had to pull a heavy box using a rope to get a food reward. The box was too heavy to pull alone and chimpanzees had to work in pairs, pulling in unison, and both would get a reward. Josh showed a film of the process, in jumpy black and white. The style felt so comedic and kitsch that you could almost create enough distance between yourself and the chimpanzee to laugh along with it. But the second film Josh showed was more difficult to swallow. In it, one chimpanzee had been food restricted, so it was hungry and highly motivated to acquire the food reward. Her partner was well fed, less motivated, dropping one end of the rope and looking away, losing focus. The first chimpanzee prodded, cajoled, encouraged her partner, and, when they were ultimately successful, quickly scoffed the reward at the end of the experiment. Forget the big plexiglass mirror in the grass outside: this was us. The chimpanzee did exactly what we would do.

Josh had to adapt the idea. He created the apparatus scaled

to elephant proportions: a sliding table that rolled forward when the elephants pulled on a rope, presenting them with a bucket and a small food treat – sunflower seeds. He couldn't make it too heavy for the elephants; these guys could easily drag a car along. Instead, the rope was looped through the equipment, so that one elephant pulling alone would just pull the rope through. Both needed to pull to get the required tension and force so they could get their treat. He couldn't keep the elephants hungry. So instead he introduced a time delay, releasing one elephant before the other so that they had to wait for their partner and pull together. Being released first and pulling would just make for that rope unlooping and no sunflower seeds to hoover up through their trunks. I stood, video camera in hand again, as Pepsi and a young female named Taengmo (watermelon) expertly executed the task, with Pepsi waiting for 40 seconds, not even touching the rope. Then they tugged together, in unison, and the rewards slid right to them. They fished out the sunflower seeds with their trunks and blasted them into their mouths. I didn't see all the hard work that went into the design or the training of the elephants, but I saw the amazing outcomes. After the session, Cherry, the veterinarian looking after the elephants, now a New Yorker herself and Josh's wife, was in high spirits as was I. Cherry climbed up on Taengmo and I climbed onto Pepsi. We rode around slowly as if we were kids learning to sit on ponies. Josh, who was getting us water, looked down and frowned at the two frivolous people playing with the elephants and prepared himself to tell them off. But when he came down to the experiment area, he realised it was us. We were too happy.

It was Josh's take on elephant thinking, the elephant perspective on the world, that shaped all of my own future thinking. I had already started to think about elephants in their environmental context, including the human interactions that they had. But how personality, risk-taking, and their sensory experience played into being an elephant was totally new to me. I always liked to measure things that I could easily quantify: births, deaths, size, hormones. Classifying behaviour is so much more difficult because it requires a very nuanced knowledge of a species: a willingness to classify things knowing one might be missing something, but thinking also that it's more valuable to do so than not to try at all. So taking what has been done before and reconsidering it. I thought the things I was interested in were entirely separate from behaviour. But I was wrong.

Yet another example from Amboseli struck me. The detailed individual-based data collected over many years came into play again. The researchers there were able to measure 650 footprints from 302 different individuals between 1976 and 2007. Footprint size is a great estimator of body size, and importantly for us science nerds, it's validated, so we know how it relates to elephants that have been measured objectively. The Amboseli footprint database included 120 measurements collected from 36 unique individuals that the researcher observed to eat crops (sometimes called 'crop raiding') between 2005 and 2007. So the Amboseli researchers had a marker of body size and information on one of the riskiest things an elephant can do, which is eating crops that humans are growing – precisely the kind of huge risk that Derek the elephant took in South Africa, and that could easily have cost him his life. By putting them together, they found that for males aged 16 and over, just the age Derek was, eating crops had a positive effect on foot size. It meant they were bigger. And individuals that started to eat crops weren't any larger for their age than one would expect – the growth came after the behaviour. This suggests that eating crops affected male body size rather than male body size determining raiding status. It made me see Derek and his elephant friends in a whole new light. Before, I'd just thought they were getting cheap calories in the form of tasty mangoes. But they might end up with a competitive advantage by having bigger bodies, and might be better equipped to win a fight like the one Soshangane died in. The risk suddenly appeared to balance better with the potential gains, and all this because I had separated out body size (something I found easy to understand) from behaviour (something that seemed too nebulous). But really, Derek isn't just a product of his age, his sex, his size; he is also a product of his behaviour, his actions, his personality.

One day, far from Thailand or South Africa or any elephants, I was having lunch in Cambridge at Pembroke College. I was chatting to Nick Davies, an unassuming and gentle man who just happens to have a towering intellect and the reputation to match it. He was pivotal in developing behavioural ecology as a discipline. But he would never make a fuss about it; he'd be more excited to show you the peregrine falcons nesting in the church tower opposite our college. He had been a fellow at the college since the 1970s. He told me that the idea of animals having personalities had seemed out of this world when a senior colleague and great thinker had mentioned it to him in the 1980s. Now it's standard, accepted as a part of individual variation. There's an exciting study on the topic in the Myanmar elephants, which reports on the characteristics oozies attribute to the elephants they work with. The real joy of the Myanmar study is the sample size: they were able to interview 316 oozies about 257 elephants. Oozies often talked about more than one elephant they knew, to give multiple perspectives on their personality. This kind of approach certainly tells us about human perspectives on elephant personality. I would argue that's important and interesting, particularly when elephants live in such close contact with humans. The study found certain traits, which the researchers asked about, clustered into three broad categories. There was 'attentiveness' (which included traits like obedience, 'slowness', vigilance, confidence and activity). Then 'sociability' (related to mischievousness, socia-bility, playfulness, friendliness, affectionateness and popularity). Finally 'aggressiveness' (constituting the traits aggressiveness, dominance and moodiness). The findings were very similar between male and female elephants. So these traits could be an elephant 'big three' similar to the 'big five' in human studies, which use survey data to describe personality in terms of openness to experience, conscientiousness, extraversion, agreeableness and neuroticism. In the elephant's case, each individual forms a different-shaped triangle along the three axes. I think I'd be some kind of isosceles. But, being human, I suppose I'd settle for my jagged pentagon.

Of course, there's also an Amboseli study on a similar topic. Amboseli is our wonderful resource, our Kilimanjaro, our great big elephant hope. In the study, observers who knew the elephants well, having studied a specific group known as the EB family, all female, rated the females for different personality characteristics. A total of 11 females whose social status, age and genetic relatedness were known, were included in the study. Rather than three clusters, the researchers found four. The first was what they called 'leadership'. This category included effectiveness, permissiveness, intelligence, insecurity (a negative relationship, with less insecure elephants having higher leadership), confidence, opportunistic personality, as well as equable, strong, and maternal characteristics. The second category was 'playfulness', which included measures of activity, curiosity, playfulness, excitability, eccentricity, sociability and slowness (a negative association). The third was called 'gentleness', comprising a strong negative association with irritability, aggression, and deferential personality traits. The final component was 'constancy', which included predictability, a negative association with fearfulness, popularity, protectiveness and sensitivity.

What I found most interesting about the study, other than being able to make quadrilaterals rather than triangles by linking the clusters together, was that the leadership cluster of personality traits also correlated with social status. And leadership was the only component with any relationship to social status. In other animals, this could be quite different. For example, in chimpanzees and orangutans, dominance is a strong predictor of leadership. In hyena, some would argue it's assertiveness. In elephants, particularly females, we know age is tightly linked to social status. But since the Amboseli elephants were so closely followed at the individual level, the researchers were able to state that social status explained more of the leadership than age seemed to, because there are slight variations in the status of individuals in an elephant group that aren't just about age. The researchers stated that their leadership cluster, rather than reflecting boldness or dominance, might actually be closest to the trait of openness in the five-factor human personality models, representing influence, knowledge, and perceptual abilities. This tells us a lot about what qualities elephants need in order to lead.

If female leadership is about effectiveness and intelligence, how does this play into the landscape that elephants navigate, one that often contains a lot of risk? And what about male leadership? A classic paper by the lofty Asian elephant expert Raman Sukumar suggested that the risk-taking behaviour of male elephants played into their propensity to do that riskiest of things – eating crops like Derek the elephant did in South Africa. Raman conducted a study thousands of miles from Derek in southern India. I was in that area in the summers of 2014 and 2018. We drank milky masala chai as the rain fell in sheets around us, then pooled at our feet. The sheer volume and force of it made the brightly painted temple, covered in the statues and reliefs of many and various gods and characters of Hindu myth, virtually obscured. I was as drenched as the dogs looking for shelter. We were in an area called Bannerghatta, just outside of Bengaluru, an ancient village on the edge of a bustling city, one specialising in technology, and also on the edge of the elephant habitat. I knew how close the elephants can come to human-occupied areas, but this was more extreme, more of an edge space than I was used to. It wasn't so much of a transition zone as an everything space: people fixing cars, cooking, farming, commuting. I'd been able to take an Uber, which crawled out of the city centre past the stone cutters and street-side restaurants, between the mopeds, to the field site. It was separated from the city more by traffic than by distance. And when I got out of the car, there were elephant fences outside smallholdings. There were NGOs working on coexist-ence between humans and elephants, and behind it all the palpable tension that comes when people try to grow their livelihood with hungry elephants around.

Back in 1988, Raman Sukumar and a colleague of his found that adult males went to eat crops more than family groups throughout the year. Between October and December, after rains of the kind that poured on me in August, an adult male on average would eat crops on 38 nights and a family group on only six nights. So adult males derived about a quarter of their total food from crops, and families less than 5 per cent. Across a whole year, it was about 9 per cent for adult males and from 1–7 per cent for family groups. Males are bigger than females, the ratio being about 1:1.14 in Asian elephant body size. Sukumar framed these differences in the context of sex differences. Think about the males eating crops in Amboseli; they could experience the benefit of increased body size. Sukumar also noted that finger millet, cultivated grasses and paddy, some of the main crops, had higher protein, sodium and calcium than wild grasses, so elephants could take in a lot of nutrition by foraging in crops. Sukumar highlighted that there could be a lot of variation in male elephant reproductive success, with some males able to father many offspring, and others outcompeted. This is in contrast with the usual situation in females. In fact, most adult females in the wild, regardless of their size, tend to reproduce (although it's certainly not the case for females in captivity). Sukumar argues that this potential gain, coupled with the fact that male elephants don't participate in parental care of calves and that they don't associate so closely with their kin, and therefore don't risk hurting their whole family if they get into a difficult situation with humans, explains the difference in the male and female strategies. Whenever I talk about this, I see this recognition of such an association between risk and males in the faces of the humans I speak to. Humans have different social structures and we often live less sexually segregated lives, and yet when I talk about those young adult males and their relationship with risk, there's a familiarity about it, but also a risk of generalising and making too many assumptions.

Then there's also risk to humans when elephants are eating crops, something which used to be referred to using the rather loaded term 'crop raiding'. Usually, the humans want the elephants to move away, so they can protect those things they have spent so much time and energy and expense growing. But chasing away a hungry and perhaps grumpy elephant can lead to injuries and even death on both sides. And sometimes people are just in the wrong place and surprise an elephant, or cross their path. Furthermore, when I write about people, it's not all people. Those of us who like elephants, do not bear the negative side of having elephants around equally. It's not equal at all. Someone who is injured or loses money from a crop-foraging elephant is unlikely to be someone who looks like me. Elephants, like most biodiversity, are concentrated in the tropics, closer to the equator, in what might be called the global south. And within that, I don't think it's any coincidence that it's people who have fewer resources and sometimes less political power who bear the very highest cost of the downside of interacting with elephants. They might be concerned about their personal safety, how they're going to get around at night, the structures they live in, the quality of the fences around the crops, the safety of their kids walking to school, and the prospect of running into elephants when herding their animals. At a Friends of Elephants meeting in Bengaluru, we discussed this with great passion, after watching a film made by the journalist and documentary-maker Jyothy Karat. The people might descend from those who were moved off land to create a protected

I never got tired of taking pictures of elephants and mahouts interacting.

An elephant shows us some timber-moving skills in Myanmar.

Thaung Sein Win, note his mahout lying down on his back. (Copyright Carly Lynsdale)

Shading the scale reader so we could see the digital figures in Myanmar.

Research on Myanmar
elephants. Process of weighing.
(Copyright Alexandre Courtiol)

Research on Myanmar elephants.
Measuring foot size. We were
using this to estimate body size,
but the elephants often didn't
want to put weight on the foot
with the tape measure around it.
(Copyright Alexandre Courtiol)

Elephants reach out their trunks for bananas we were giving out after taking blood samples in Myanmar.

The smaller elephant is comforting her mother while she has her blood taken by a veterinarian in Myanmar.

A big male working
elephant in
Myanmar. His tusks
have been trimmed
to prevent injuries.

Pepsi and his mahout. Pepsi has been trained to pick up the flip-flops of tourists and his mahout has tied them around Pepsi's neck.

Baby Carly (the smallest elephant) with her family.

An elephant reaches for my camera.

A white elephant calf in Myanmar.

area or a tea plantation, and now the elephants are there, either because it's been designated a space for wildlife or because the elephants are attracted to the area. Sometimes, I think it's even harder to see yourself in another human being than in an elephant. Because that human can be so different to me: their life, their experience, their language, their perspective. People can make choices I don't like, and maybe we even have different ideas about what can be chosen in life. But if you care about elephants, I would say, please, *see* other people. See the immigrant Sri Lankan labourer whose story Jyothy captured, who moved to southern India to earn a meagre living on the tea plantations, who worked as a janitor, who had to walk early to work. Who walked straight into the path of elephants that were being frightened away from crops by forest officials, themselves trying to ensure the safety of the public and the animals. Despite that, the janitor never arrived to work or saw his children again. Instead, his wife wept in grief by his funeral pyre. I don't see bad guys in that story, only victims, human and elephant. The facts tell us that 1,144 people were killed between April 2014 and May 2017 in India by wildlife, mainly by elephants – that's almost 400 every year. Every single one of those people has a story just as unique as the stories of Baby Hannah, Soshangane, Pepsi and Yeager.

When I talked to Josh about this, I was sad. And the great thing about Josh is that he knows something about empathy. He'd even written about it in elephants. In one study, he had observed 26 elephants for one to two weeks a month for almost a year. He was interested in how elephants reacted to another elephant having what's called a 'distress event', which is basically when an elephant experiences a negative stimulus, such as the threat of being separated from their group, or an aggressive signal from another elephant. That stimulus drives an elephant to react; it becomes agitated and it signals this agitation to others in its group. These signals can be changes in its body, like tilting its ears forward or its tail sticking out erect; it can move and it can vocalise, for example with a trumpet, roar or rumble. Josh found the reaction wasn't limited to the elephant experiencing the distress event; rather, there was a social contagion effect, with surrounding elephants reacting in the same way. This process has been used to suggest empathy. The elephants also displayed increased affiliative behaviour to the original elephant in the time immediately after the distress event, including physical contact, such as trunk touches to the genitals, mouth and the head, and vocalisations. Josh found that the high-pitched 'chirp' vocalisation (the one I usually think of as a squeak) was emitted by the elephants more than usual following the focal elephant experiencing distress, and they also did more 'trunk bounces', when they audibly bounce their trunks off the ground. This is all interesting, but could be delayed reactions to the elephant who first reacted. However, Josh did note that the elephants directed their behaviour to ward the elephant who just reacted, and in this way, the behaviour has a lot in common with consolation in primates.

In addition to vocalisations elephants use their incredible

olfactory system to engage with the world. You can see this system if you just look at an elephant skull. Inside once sat their brain: weighing up to 5.5 kilogrammes in Asian elephants and 6.5 kilogrammes in African elephants, several times the size of the 1.5 kilogramme human brain. It makes sense that the biggest land mammal has the biggest brain. But from the outside of the skull, the first thing you'll notice is there's a large hole at the front, which might make you understand where the myth of Cyclops comes from, particularly given that dwarf elephants used to roam Mediterranean islands like Sicily, Malta, Cyprus and Greece (in fact all major islands excluding Corsica and the Balearics have fossil evidence for them, dating from the Pleistocene). This hole is actually for smell, not vision, and it's where the trunk meets the head. What sits there is not one huge eye, but two rather sizeable olfactory bulbs. Having these structures in the forebrain that receive input from the nasal cavity is not unusual. Mammals have it, as well as reptiles and other classes of animals. What is incredible in the elephants is that they are so big, even for the size of the animal. We know that elephants have almost 2,000 functional genes dedicated to scent: one study found precisely 1,948. This compares to 811 in dogs, 1,207 in rats, 309 in macaques and 396 in humans. The elephant world might well be a smelly one. I remember walking past a large male elephant with his oozie perched on his head on one of my first trips to Myanmar. The elephant reached out his trunk to me as I walked several metres in front of him. He seemed calm, not freezing as elephants sometimes do in groups when they are processing some information, like sounds and scents, from far away. But his trunk was on the move. I reached out to him, but he did not touch my hand as I'd expected. He sniffed, staying 30 centimetres or so away, and then he retracted his trunk. At the time, I thought this was odd; he seemed to have wanted to touch me and then changed his mind, but perhaps that sniff was all he wanted.

I know elephants can detect a lot from those smells. In South Africa, I sat up on the back of the *bakkie*, in the open air. It was a spring morning and we were driving among the mopane, winding along a dirt road near the river that is named after the elephants, the Oliphants. We stopped ahead of a group of elephants that was about to cross in front of us. It's sometimes better to do it that way, for them to come to you. We were like a rock in the river, elephants passing either side, calves with their mothers, walking close, older females and more curious young elephants. Some fanned out their ears, showing they were aware of us. And at the back, a few males followed; perhaps a female further ahead in the herd was in oestrus and they were tracking her. A young male, maybe about 20 years old, was interested in our vehicle. He stood in the road and stretched his trunk out, sniffing. But it wasn't just a sniff this time. He walked over, calm and steady, as my heart started to race a little. He held out his trunk, ridged with 50,000 rings of muscle, and he sniffed. He came further forward. I looked down and tried to breathe steady, fearing any sudden movements would startle him. But I was finding it hard to stay calm. His trunk was not like a hosepipe, it was thick and heavy. He tilted up the brim of my hat at the back, a wide and leathery old one, and sniffed again. I felt the hairs on his trunk through the hairs on my neck, but I didn't feel his skin. I tried to remember to breathe. I wondered if he would break my neck. He was still sniffing, but then, he stopped, turned, and walked back, joining the flow and those following the herd. I was a curiosity for a moment, and then I wasn't. But my heart still raced. I knew now what a hair's breadth meant.

On my second visit to northern Thailand in 2012, things were similar to the first time and yet different. It was balmy and less wet, towards the end of the year, yet still green from the rains and not scorched with heat. And this time, I took my sister, so she could see some of the wonderful experiments, so we could eat on the little plastic stools by the river and take a boat to Laos. So she could meet everyone. The same cast of characters greeted me: Pepsi, even bigger this time and looking like the male he was going to become; a big male named Phuki, with pretty tusks; and lots of females too, such as the smart and playful Ploy, who always gave you the sense she was capable of anything, as long as it was in good fun. The science had moved on too and the mirror and cooperation experiments were no longer the focus. Josh was working on an olfaction experiment.

Josh and the team still had that sliding table mechanism from the cooperation experiment. But this time the buckets were covered with lids. When my sister and I helped out early one morning, the elephants were well into the experiments and knew what was going on. But I was learning. We watched a female named Bleum tackle the experiment first. The door was pushed forward and she was given the chance to sniff both of the buckets she was presented with. They were plastic, with lids held on by cable ties, but the lids had holes, so she could sniff through. Bleum touched each lid delicately with her trunk, and took a few sniffs. She rested the end of her thick trunk above them for a few seconds; it seemed as though she was processing. Then she stepped back. Leisurely. The researchers slid the buckets back, away from Bleum, and cut the cable ties. Then they offered the buckets back to Bleum and she chose one, sucking up the sunflower seeds that were at the bottom with her trunk and blowing them out into her mouth. At first, I thought, well, fine, she can smell which bucket contains the seeds. But it was more than that, Josh explained to me; both buckets had seeds, just in different ratios and the elephants consistently chose the bucket with more seeds. Effectively, just from a sniff, they could select the bucket with more seeds. I like data. And when I plotted these data, it was perfect. No fancy statistics required, a perfect sliding scale: the more similar the quantities in the buckets, the more trouble elephants had in telling them apart, but the more the buckets diverged, the more successful the elephants were, and successful at a level way beyond what we might expect. Some of the elephants reached 100 per cent. That is some nose.

Josh and I talked a lot. We both tend to. But between the subjects of families, jazz and the sights of Berlin, we talked an awful lot about animals, especially elephants. We both really believed that what we study – the lives of elephants, the minds of elephants, the elephant experience as it is born and lives and dies – is fascinating. It's also useful. With those olfaction experiments, Josh was really onto something. Something that hinted at a shift. We thought about how it linked with other issues, such as elephants eating crops. Oftentimes, it's easy to frame this as the elephants doing something that destabilises human livelihoods; or to take the opposite stance, that humans living in the same areas as elephants upsets the balance of ecosystems, that humans encroach on elephant habitat. But those neat little narratives don't work. Humans and elephants are too complicated for that. Perhaps it would be possible to take both sides into account in the future. So that there aren't just deterrents for elephants, but also a kind of thinking that includes making attraction and deterrent specific to them, understanding them as individuals and knowing that we can't just scare them away. We really have to decide whether we want elephants around and, if so, where we want them to be.

As I watched Bleum, Pepsi and the others sniff and choose in Thailand, never tiring of seeing them sucking up their reward from the buckets, I thought about elephants all over the world, about their sense of smell being powerful enough to distinguish between small quantities of food, between humans from different populations, between each other. They are probably able to pick up much more than this from skin, clothes, tracks written in the wind. We, such visual creatures, aren't like elephants at all. We *see* the world, much more than we sniff it. We know what it's like to depend on a sense that way and perceive the world through it. It's even more fascinating in some ways that elephants do it differently. And somehow I know that if we can take that understanding and apply it, see it from the side of the great nose, we can do much better for them. Maybe even make a world we can both live in. Some of Josh's experiments illustrated the difference between what we think we know of elephants and what they really do perfectly. In another experiment, the task was simple; elephants have free choice to pick a bucket out of two from which to eat sunflower seeds. But only one bucket contains food and the only cue that they're given is their oozie pointing at the bucket with food in. I think most humans would recognise this cue. If they trusted another person, they would follow it. They might even expect their dogs to follow it, but would an elephant? The answer is, Josh didn't find any evidence for them following the pointed finger beyond chance. Other studies have found such evidence in African elephants, so I wouldn't rule it out completely, but the signal doesn't translate well. And that's something we have to be aware of. We can see how we are like elephants, but part of this requires seeing how they're also not the same. It's kind of like Pepsi and the mirror. First, it's a plexiglass one, which is never going to give you the perfect reflection. Then there are these other barriers, the differences in sensory modalities, in grooming habits, in how likely one is to touch a mark. And the fact that Pepsi still passed the mirror test in spite of all those things, I still find it amazing.

CHAPTER 8

Listen

*Some people talk to animals. Not many listen, though.
That's the problem.*

A. A. Milne

I talked to elephants a lot. In Myanmar, I learned how oozies tell them to stop, to turn, to lie down, to push, to pick up. Things you might want a huge machine for manoeuvring logs to do on command. And the elephants even listened to me: when I said 'stop', they responded, pausing for me to take a photograph. Their oozie held down their ear so I could capture any tears, holes or jagged ends, the physical manifestations of their lives in the forest that would help me to identify them. Just as we all have our own tears and scars, but they're always different. One hot and dry day in Myanmar, we kicked up the dust driving to a remote camp from Katha. We bumped along the sandy roads for over an hour and I laughed that we'd only travelled a few kilometres when I looked at the GPS. Hannah, I thought to myself, you have to measure distance in time, not kilometres. We'd set off fairly early, after our sweet tea, coffee that comes in a sachet with milk powder, chickpeas, paratha, and all the fried things I love for breakfast. The sun was high and hot by the time we arrived at the camp. I shielded my eyes with my folder of documents. The oozies were dressed in bright yellow tracksuits that stood out like neon highlighters in the sun. We organised our workstations: space for blood samples, digging a hole to embed the elephant weighing scale, a production line to weigh out dung samples, some of which we'd dry out to extract hormone metabolites. I wanted to know how glucocorticoids, ones linked to 'stress' response, varied across seasons, between individuals, in the different sexes and age groups. They are part of that fight or flight response that can be healthy or can break down, become chronic and weigh heavily on these huge animals, just as it can for us. A drop of sweat dripped from my brow and I turned away to avoided getting it into the dung I was working with. If I were really an elephant, I'd only sweat between my toenails. But here I was, a human, and I did not want to contaminate the sample. I didn't want to know how stressed I was, and I certainly didn't want that getting in the way of my studies.

You see, that's how I thought the elephants spoke to me. I could do it directly to them, and in return, I did these things, procedures, sampling, statistics, to try to hear back from them. And it's funny because in terms of my experience, I already wanted to do more than that. At that camp, I took photographs of the elephants, sometimes straying from the ones I used for identification and analysing size, where the elephant had to stand still and perpendicular to the camera. I turned and caught the oozie in a plastic helmet sitting nonchalantly on top of his young female elephant. She chirruped excitedly, reaching out her trunk for a banana treat that one of the other researchers was proffering in thanks for a blood sample. I also noticed the oozie's foot, hooked carefully below the elephant's ear, making sure, even with her excitement, she didn't move forward too fast. When another female elephant had given her blood sample, a younger female elephant stood next to her and put her trunk by the first elephant's mouth; she sniffed around and even placed the trunk in the mouth of the other elephant. They rumbled back and forth to each other. It was a gesture of comfort, the elephant making herself vulnerable by putting her precious trunk into the mouth of another, and with a contact call at the same time. I captured the trunk movements with my camera, not wanting to forget them, but the sounds were also with me.

When I returned from the field, I clicked through the brightly lit series of photographs. Between all of the elephant pictures, there was one of me standing in a line with girls in their longyis, all of us with little yellow flowers in our hair. I was a head taller than the rest, but we all held hands and grinned. There was another of me turning to smile as we walked up the steps to a temple. Effectively, they had become holiday snaps. I went to the laboratory and sat in front of my computer screen, pulling up the result from stress hormone analysis on my screen and tracking monthly varia-tion in hormones first with my eyes, then with my tests and models. I was looking for peaks and troughs, for that worrying constant elevation. I was pleased not to find it. In consecutive months, the levels we measured might rise, perhaps linked to high workload in the monsoon season, or other seasonal factors. Maybe the abundant food in that same season could also be linked to the spike, because eating, digesting, metab-olising and all that activity could be associated with hormonal changes. Months later, I sat back from my written-up paper, knowing I had made a contribution, but that I had also opened up more questions and more possibilities. I was happy to let the data speak to me and listen to it. But I hadn't really listened to an elephant yet, and I really wanted to next time. I clicked through the photographs again and thought I was missing a trick. What could they tell me beyond me pulling data out of them? What did they just say to each other?

So I decided next time I would listen. And I took it quite literally. There are many ways elephants communicate, through touch and scent and sometimes peeing right where they stood. I know of an elephant in Nepal that is not fond of male elephants at all and pees whenever one walks past her. But I thought literally listening would force me to consider all of the communication. So, almost two years later, after I had performed and cajoled and convinced people to fund a study, I sat with field assistant Jess on the back of a pickup truck in South Africa. We shared an umbrella for a bit of shade. We had a microphone encased in a special windshield with a grey fluffy cover. I noticed it looked monstrously close to a skinned version of one of my mother's cats. But we were miles from my mum's garden, where Mpumalanga meets Limpopo, and I was meant to be present and listening, not up in my head. Jess carefully adjusted the settings of our recorder and I fixed a pair of headphones over her head. I put my thumb up at her, but I didn't fall into my usual chatter pattern. I had to shut up and listen to the elephants. So I looked beyond Jess to six male elephants standing in a shallow, wide pool. Come on boys, now's the time, talk to us. Four hours later, we had recorded one faint rumble, and it possibly had a layer of wind over it, making the sound file unusable for further analysis. I had retreated into the vehicle and Jess toughed it out alone under the umbrella. It was at least 35 °C. So this is why it seemed as though people hadn't been recording male elephants much. Maybe they didn't talk that much, and when they did, the wind blew it all away.

Both African and Asian elephants vocalise. They do so in many different circumstances, and the sounds they make vary too. For example, there are rumbles, the most commonly recorded vocalisation. These are laryngeal and have a low frequency, often with their lowest elements falling below the human hearing range. This makes them 'infrasonic' to us, but to the elephants they are sonic, because they are perfectly capable of hearing them. We call them rumbles because they sound that way, often falling somewhere between a deep tummy rumble and the engine of a faraway vehicle. (I say this because I've mistaken both of these for elephant rumbles, and, after the former, had the field team turn and stare at me accusingly.) They might be used to contact other elephants, or to communicate movement patterns. But it's not that simple: there is also the pulsating metallic musth rumble that is associated with that specific hormonal and behavioural state. And rumbles can be used in combination with other vocalisations, such as a roar, which is higher pitched, and honestly more 'rarrrrrrrrr'. There's a reason it's called a roar: lions certainly aren't the only ones doing it.

And then there are the other vocalisations, less common, but relevant. The incredible resource Elephant Voices lists all of those that are known for African savannah elephants. The laryngeal ones include the rumble, rev, roar (the one I know, with subtypes of noisy, tonal and mixed), cry, bark, grunt and husky cry. And then there are the trunk call types: trumpet, nasal-trumpet and snort. The database also describes them and gives examples. I played them and tried to emulate the sounds. My 'trumpet' was particularly weak. Luckily humans are good at making things and we can imitate it with a musical instrument. The elephant trumpets can signal excitement and be part of play or can be linked to nervous excitement, perhaps something unexpected happening. But where's that squeaky chirp I heard so often in Myanmar? Well, African elephants don't tend to make it at all. In fact, the only case of one doing so was a male who had been living in a zoo with Asian elephants for years and probably picked it up from them. And some vocalisations, such as the 'cry', are only made by infants, often wanting to draw attention to their needs, rather like a human infant. Some are imitative; for example, the elephant in a Korean zoo, which copies the voice of his handler, or the sounds that some elephants in zoos make that sound like engines starting or motorbikes. There are social, sex-based, age-based, species- and context-specific elements to these sounds. And there I was just huddled with Jess under an umbrella, trying to catch one of them at a time!

The variety of calls and contexts was off the charts interesting. Collecting that kind of data was brand new to me. Hearing the buzz of insects, the call of a bird and sometimes, just sometimes, the crystal-clear low and sonorous rumble of an elephant through those headphones, everything heightened, was a powerful experience, although it could also make one jumpy. Everything sounded so close that I developed an aversion to being tapped on the shoulder. I could even smell better when I had the headphones on. I separated scent of the ground from the dung, from the sweat, from the insect repellent I was wearing, and saw all those layers just like the harmonics of a rumble, together, at the same time, making something complete and real. When Jess and my other assistant Amy plotted out some of those rumbles, they chose colours of white, purple and red on a black background. Each rumble looked like stacked layers, like a cake. Those are the colours I see when I now hear a rumble, like lines of fire and ice against the darkest sky; elephant volcanoes, bubbling up and reverberating through the earth.

One day we were looking for Sizwe, an older and spectacular

male. Some days we would drive the dirt roads around and around, circling him as he ate deep inside a block of vegetation. The bleep from his tracking collar would always be coming from the same spot, out of reach. After a couple of hours we would give up and have a break, a sandwich, wonder if he would move or we should just try to find another elephant. But that day he was out in the open, drinking by a shallow pool. My assistant Amy had gone out to South Africa ahead of me and struggled with the winds and getting the recording protocol just right, and typically, I showed up to this. Almost no wind, four other males and a back and forth of rumbling between them that is so rare. Amy carefully adjusted the recording settings so that we could make more of those beautiful spectrographs and pull data out of them too. I gave her my widest smile, tried not to talk and we just listened. We could have sat there for hours. But we are not the only ones interested in elephants. People on their holidays, for example, might stop by in their vehicles and want a look and a listen too. Our policy is always to make space for this and to be as unobtrusive as possible. So that day, even though the data coming through was incredible, we backed off a bit and gave people in other vehicles the chance to enjoy the elephants too. I leaned back in the vehicle and stretched out my legs, wondering how many hours I had spent watching elephants, trying to see them and hear them, trying to make sense of their lives. Even if some people just want ten minutes and some nice photographs, they should have that. It's hardly my place to monopolise them. And anyway, today Sizwe was not interested in us humans. I watched him spray water in his mouth and rumble again, deep and not at all growly. He would be moving on soon, no longer a spectacle or data source, just an elephant.

It was my typical approach to research. Some people seem to be able to be so focused and to be able to deconstruct things into smaller and smaller parts, making a reductive scientific Russian doll. Everything I learn always seems to make things bigger and more connected for me. The sound, the action, the social association linked to it was another layer of complexity in a huge elephant world that I knew a little bit about but not enough. I remember a senior colleague looking at me sceptically over the top of his glasses and saying, 'Hannah, that's a lot of moving parts.' That's scientific code for: it's too much, too many unknowns, too likely to go wrong, too overwhelming. I barely even heard the thinly veiled criticism; all I could think of was an elephant made of cogs and wheels and the parts all moving within to make the whole thing move and rumble and shake. I put an image of the cog-work elephant in the corner of all my presentation slides, to remind me that the cogs are interesting and the whole individual is too, as well as all the others and their respective places in the world.

I read about a biologist long before inviting her to meet me in Berlin. She was called Caitlin O'Connell Rodwell. She thought big in a really interesting way. The vocalisations of elephants had been studied before, and classification tools had been developed to organise them in our scientific minds. Caitlin and her collaborators went beyond what we could hear of them and into the infrasonic components of elephant vocalisations and also the stomping of their feet. She and other scientists investigated the fact that elephants could communicate using the infrasonic elements that can travel through the ground, contacting each other up to 10 kilometres apart. And there's also the part that travels through the air. There are waves that travel through the ground and elephants don't pick them up with their ears, but with their feet and trunks. And those waves can travel 16 to 32 kilometres. It's like elephants being able to see beyond the horizon. Long before we had the internet or phone or maybe even smoke signals, the elephants were part of a bigger, more connected world.

To understand how it works, I remember all the times I've described the structure of an elephant foot, holding up the bones and saying, 'An elephant is up on its toes, but the body weight piled onto the foot is supported by a big fat pad, back here.' I wave my hand under the elevated heel. 'This allows them to support their body weight and to make those nice, silent padded movements through the landscape.' But I also remember to say that the foot isn't just about locomotion, it's a device for picking up signals that vibrate through the ground. Caitlin discovered that elephants have enlarged ear bones as well as sensitive nerve endings in their feet and trunks that might be associated with infrasonic communication. She suggests that seismic waves could travel up the toenails to the ear through bone conduction, or that somato-sensory receptors in the foot similar to ones found in the trunk could be used, or perhaps it's a combination of both. Those powerful receptors help to explain why elephants could pick up the sound of storms from tens of miles away. Caitlin said that at her study site in Etosha, Namibia, elephants started moving north towards Angola, 160 kilometres away, when it rained there. Perhaps they could detect the thunder.

She also mentioned that memory and the reaction to trauma were caught up in the soundscape of elephants. For example, the 'thuk, thuk, thuk, thuk' whirring of helicopter blades could cause agitation in elephants, particularly if they associated them with something negative. In elephants that lived through the culling years in South Africa, they could react with fear to the sound of those blades, associating it with distress. And these kinds of reactions weren't limited to anthropogenic noise, but included the sounds they made to each other. Caitlin documented an experiment in which she played low-frequency 'alarm' vocalisations emitted by elephants to eight captive individuals at a tourist site in Zimbabwe. She used elephant calls, synthesised low-frequency tones, rock music and silence to compare the reactions of the elephants to those stimuli and the elephant warnings. Consistently, when the warning call was played into the air, one female became extremely agitated; she even leaned down and bit the ground. Caitlin described it as very unusual, but an observable behaviour that happened in the wild. There was something specific about the content of that alarm sound, not just the presence of any low-frequency stimulus that affected the elephant. She observed reactions in the other elephants, all male, too, although these reactions were less dramatic. But males don't live in family groups their whole lives, so perhaps that contributed to the difference, or maybe the female had had a particular experience in her past that led to her stronger reaction.

I very much wanted to meet Caitlin, I was so excited by her testing of some really huge ideas. And when I did, I wasn't disappointed. She was every bit as knowledgeable as I had imagined, but also very calm, pragmatic and very generous with her time and advice. In my opinion Caitlin had shown both creative thinking and an extraordinary commitment to studying carefully and closely. We sat in an airy room in Berlin. And Caitlin captured everyone in it. She described how she and her team recorded observations of male elephants 24 hours a day at a waterhole in Etosha. Repeating this every year meant the study reached a fantastic level of resolution for years and then decades. She showed a picture of the observation tower she'd built with her team and played a time-lapse video: data was being collected day and night. Oftentimes, we just watch and listen to elephants during the day, and at night we rely on tracking technology, but with the help of infrared cameras Caitlin captured interactions over the entire daily cycle. And that's a really important thing, because elephants don't sleep as much as we do. In captivity, it's four to six hours a night, and one study of wild savannah elephants in Botswana found they only slept two hours a night. They only had REM sleep, the kind in which we humans dream, every three or four days. That's one way I massively fail at being an elephant: I need a lot of sleep. This is a particular issue when I do fieldwork and my sleep is affected by medication. When I take anti-malarials, I dream I'm falling into a Van Gogh painting with 'Lucy in the Sky with Diamonds' playing in the background. After dragging myself out of a swirling 1960s take on 'Starry Night', I have replied to 'Good morning!' with a perplexed expression and, 'where's my hat/tea/bag/biscuit?' more than once.

I remember listening to the renowned conservation scientist Kevin Gaston talk about what he called 'the ecology of the night'. I liked the sound of it. It felt somehow subversive and dark, beyond there just not being as much light around. Kevin explained that perhaps our human constraint of being active in the day had potentially not only led us to disproportionately study animals that shared that activity pattern, but also to fail to acknowledge how much our love of light, and night-time light, might have affected all of that other life. I thought about elephants. It was another way I wasn't listening to them. But luckily, someone else was. Researchers from the Elephant Listening Project know that oftentimes forest elephants are not seen, but they can be heard. So instead, they harness the fact that they can be passively recorded in order to monitor populations of those rarest of elephants. In the Nouabalé-Ndoki National Park in the Republic of Congo, the Project is setting up a grid of acoustic units, covering 1,300 square kilometres of forest. It will record not just elephant vocalisations, but also gunshots. Which is scary but desperately needed data. It's not proposed as an alternative to traditional patrol-based methods, but to get a better handle on where elephants are and where illegal killing is happening, because of course both of these are dynamic. Then those vital patrols could be better directed to the areas of most need. The thing I find most exceptional about what the team is doing is they're not just saying yes, there are elephants here, they're interpreting the call rate they measure to determine how many elephants there are. They can use observations from elephants in clearings called bais, openings in the forest where they are visible, and extrapolate this call rate to estimate numbers of elephants from the passive recording outside of the bais. This is important for the forest elephants and potentially for other forest-dwelling animals too, whose status might have been underestimated or ignored in the past just because it was so difficult to count them. It would be terrible if we lost things we didn't even know about yet, just because we couldn't listen.

When writing manuscripts, I always call elephants by their individual names: Bulumko is Bulumko and could never be Intwandamela or Hannah or Pepsi. Most of these names, I don't choose, and we can argue about who should be able to name an elephant, but I have always preferred the names to numbers. Partly because remembering chains of numbers gets confusing and way too error-prone when you're typing. I baulk at referring to them as 'it', although I would never want to subject them to human concepts of gender. I know, however, that if I write their names I will be corrected, as 'it' is the standard in scientific papers. Just as they are still called 'subjects' in behavioural experiments, decades after we stopped subjecting humans to experiments and started inviting them to participate. For now, we can't do this with animals, and writing that elephants are participants is considered to be too personal. As if we didn't have a personal relationship. I pick through their poo and agonise about their social lives and health and what they're eating, whether it's enough, whether their attempts at reproduction are successful. That's more than I do for a lot of my friends.

But I do know that the names I use for elephants are different to how they might see each other. And I wonder how they tell each other apart. Angela Stoeger and her team in Vienna have been studying the acoustic communication of elephants in fine detail. I was extremely excited to read a paper in which they focused on males like Bulumko, who, although blind, is perfectly capable of displacing other males and therefore has some dominance in male society, and Intwandamela, with his short legs and broad body, who looks like a bulldog of an elephant. Angela and her colleagues showed that these males have variation in their rumble vocalisations. An important part of that variation is by age, which is linked to size. But even when controlling for this, there are differences between individuals. I suppose one way to interpret the data would be that everyone's voice might deepen as they get older, but you can still hear differences between different people. I was thrilled by this finding, and Angela also highlighted that the fundamental frequency, the lowest component that is often infrasonic, holds an important part of that variation. It's possible that there's more variation, not just between individuals and elephants of different sexes and ages who are affected by their physiology. What if there were also variation between regions? If you could distinguish a 'dialect' or regional difference in some of these vocalisations? It would be incredible. It would mean that things that we rely on genetics for, such as telling us where an elephant is from, could open up. And with it, even more questions, like whether male elephants picked up the 'dialect' of their mother or of the other males they got to know as adults. And what of elephants who are transported to other areas? Would they change their dialect? It's one of those fascinating opportunities that really rely on us just having the time and opportunity and technology to listen in, before our own noise drowns out everything.

Again, it's something we can measure but can't hear with our naked ears that could potentially be meaningful for elephants. I looked at a spectrograph, that 'fire and ice' colour-schemed visualisation of a rumble on my computer screen, of a recording of Intwandamela rumbling back and forth with another male. The parallel lines show the deep fundamental and associated harmonics as yellow and red lines on a black background, but they are cross-hatched with vertical lines, turning the normal, many-layered 'sandwich' image of a rumble into a waffle. I played the rumble back on my headphones, and heard 'thud, thud, thud, thud, thud'. I smelt a sticky-sweetish farmyard smell in my mind's nose. Ah yes, I remembered, he always does that; he rumbles as he poos. I grinned as I thought about earnestly telling my students to record the context of all behaviour. Context is key: film it, photograph it. You know, what you are hearing or seeing or smelling could all be connected. I looked at the datasheet I had carefully designed, little boxes to distil all of the complexity into something that could be quantified and analysed. My options for his activity included 'resting' and 'feeding'. I wrote in the notes section on my computer, 'shitting, again!'

What always appealed to me about elephants was the details of their lives, and I wondered how much of what they did they were born with, and how much they grew to know through their experience with their families. I thought of an elephant that I once tried to listen to in a zoo. I had wanted to test the recording equipment before lugging it all out to South Africa, and it seemed to make sense to test it before-hand. We set up the microphone and fluffy 'dead cat' cover, linked it up to the recorder and headphones, and waited. We waited all day. My assistant went for a coffee and I took my turn holding the mic. I grudgingly took a break for lunch, then we swapped again while he had a cigarette. The day was becoming more like a list of our breaks and ablutions than one of any contribution from the elephants. I was rather glum. There were two elephants in the enclosure and another they could see. Why didn't they rumble to each other? Then, at about 2.45 pm, the male elephant perked up. He stood by the gate of his enclosure and vocalised. I'm not sure it was a pure rumble, but it had rumble elements, perhaps combined with others. But we were excited and jolted into action and carefully filled in the form, making sure we were recording properly. Our excitement seemed to rise along with that of the elephant and he let out a series of vocalisations. At 3 pm, a large mound of hay was brought into his enclosure and he got down to eating it. He didn't make any more rumbles or other vocalisations, and none in a social context. My assistant stayed the whole week and this happened consistently. I still don't know what to make of it entirely, but it does seem clear that elephants might be using their vocal communication in very different ways in different environments. I'm glad I took the time to listen to him.

I also wondered, what do elephants hear when they listen to us? I was sure they would have got into the habit of listening long before us. As well as playbacks of animal predators, Karen McComb and the team in Amboseli have also played human voices to elephants. It makes sense. We know humans kill elephants, but not all humans. In Amboseli, women and children are much less likely than men to kill an elephant. And people who spear an elephant are more likely to be Maasai-speakers than Kamba, because Maasai speakers are usually involved in herding and pastoralist activities and come into contact with elephants through that. The situation sometimes becomes difficult, concerning access to water and grazing spaces, and elephants have been speared, particularly in the form of 'retaliatory killing' when Maasai lives have been lost because of an elephant. The agriculturalist Kamba experience less of these negative interactions because of their different land use. In a very simplified way, one might predict that Maasai-speaking men would evoke the most defensive reactions from elephants. So the researchers essentially replicated their predator playback experiments to family groups of elephants, females and their offspring. The researchers were looking for that tell-tale defensive bunching in response to humans of different ages saying, 'Look, look, a group of elephants is coming' in a relaxed and clear manner, in their first language, either Maasai or Kamba. The results in terms of the bunching and sniffing behaviour played out very much along those predictions: the elephants had significantly higher probability of defensive bunching and investigative smelling following the playbacks of Maasai voices compared to Kamba voices. In addition, these responses were specific to the sex and age of the Maasai voices. Those of women and children, the groups that were predicted to present a lower threat, were significantly less likely to produce those investigative and defensive behavioural responses. It mirrored the earlier findings of the researchers that elephants reacted to red clothing, the colour often worn by Maasai, and the scent of clothes worn by Maasai, whatever the colour. However uncomfortable the knowledge that elephants can make such fine-scaled distinctions about our sex, age, and even ethnicity might make us, I don't think it's something we should shy away from. All the senses were providing information about humans to the elephants. We were talking to them without even realising, and perhaps it was much more than our words that we were communicating. We were telling them who we were.

All the scientists who listened to elephants, rather than

just talking to them, captured more than I could have imagined. They also often used creative thinking, for example exploiting the little data that can be collected from calls in the forest. Caitlin and others said many animals use vibrations to communicate, so why not elephants? And that was so impressive to me: she didn't limit them to fit around what we thought they should be, because they are always more. So the next time I went to record the elephants' vocalisation, I thought about Caitlin. I tried to slow down my breathing and my racing heart and wanted just to listen. We parked up again by a waterhole, a deeper one this time, where Sizwe, the most majestic of the males in the area, was hanging out with several younger males. It was winter, the drier season, and some low trees encircled the waterhole where the elephants were sucking up water through their trunks, blowing it out into their mouths and dousing and speckling their wrinkled skin with splatters of mud. Then they stopped. And for a moment I could not hear a thing, not even the breeze, not even a distant engine. All the elephants froze. Some leaned forward slightly, a few raised up a leg, several beats passed, and they were still frozen. I realised I had been holding my breath. Then, as if there was a signal, they moved again, drinking, splashing mud that dripped down their torsos in thick chocolatey coats. I exhaled. Caitlin, if she had been there, might have known they had heard or felt something far away, something that didn't even register for me, and however hard I listened I couldn't hear it. But by watching and trying to listen, at least I had noticed them noticing it.

CHAPTER 9

Growing old

The world is changed because you are made of ivory and gold. The curves of your lips rewrite history.

Oscar Wilde

Win Oo had a lifetime written in his cross-hatched face and an infection in the base of his foot. He had worked in the forests of Sagaing, Myanmar since he was young. Since it was a different place. Now, or the 2014 version of now, we spray bright blue antiseptic on his footpad, and he redistributes his enormous weight. He will be cared for. Some elephants I know die in a hail of bullets, or in the climax of a fight, like Soshangane. But it takes a lot to kill an elephant; just ask George Orwell. Sometimes, hearing their breathing become slow and ragged is more painful than seeing them sag and disappear quickly. And many, like Win Oo, fade stoically with age. Elephants can look old even when they are brand new, thanks to their grey, wrinkled skin. When it's dry and dusty, you think they could be a thousand years old. Even when their skin is slick and wet, they could be an ancient sea monster. Sometimes, they seem to age backwards.

I've come to find smoothness more alarming on an elephant than wrinkles: that tight smoothness could be obesity, impaction, oedema. With Win Oo, it was the foot that scared me. It was like a crater slicked with algae from that antiseptic spray, a murky Atlantis that has no business being attached to an elephant we all loved. Lameness in an elephant is like lameness in most quadrupeds: they need to stand to live. Win Oo got old and tired, and eventually he got sick. Win Oo was never alone. He was with us, and then, slowly, he wasn't.

The pacing of an elephant life is so familiar as to almost be mundane: sexual maturity at ages ten to fifteen, and in the Myanmar elephants first pregnancies for females around eighteen. Just like humans. Offspring being dependent, suckling. The next pregnancy. The prospect of becoming a grandmother, grandparental care. Life stretching out fifty, sixty, up to over eighty years. It's a life we recognise because it's our life. As individuals, things might be different. Win Oo wasn't the oldest elephant I knew, and yet he faded. Baby Hannah's mum had many babies, and some of her friends had none at all. But the general pattern, we know it. We don't require any leaps of imagination to understand it because we are all on that path ourselves.

My grandparents have always been older, that's no revelation. But I suppose now I have to accept that they are old, in their mid-eighties. They had, for many years, a wooden elephant head on the back of their front door, carved in an orange-brown wood and shaped like a shield, with two short, blunt tusks and almond eyes that don't need pupils. It had a 1970s minimalist air about it. Now, it lies on my desk, retired from many years of looking into a family home.

I went in the door it guarded for the first time before I was even aware of being myself; I was carried. My grandparents and parents knew me before I really knew I was alive, before I could remember anything for any substantive time. They called me by my name before I knew what a name was. Their names were among the first I knew, personalised versions of the bland English forms: Grandstar (who had always been Star to my mother), Grandfarth. I went through the door, under the elephant, so many times. I played in the garden by the willow tree with my sisters, sat on the swing chair, ate mountainous quantities of food. I jumped up and down in the loo, trying to reach the string to switch on the light, imagining that one day I would be tall enough to reach it. The end of the string now hangs by my hip.

My grandparents were, are, extremely important to me. That's also the case for elephants. Mother presence, particularly in the early years, is vital to calf survival. But we also know that in the Myanmar elephants, those that have a living grandmother, and more importantly a grandparent living nearby, do better than others. A calf born to a young mother (younger than 20 years old) had eight times lower mortality risk if the grandmother lived in the same place as her grandcalf compared to calves with grandmothers residing elsewhere. Having a grandmother also decreased a daughter's inter-birth intervals by one year, so she could reproduce faster. This was all happening independently of the grandmother's own repro-duction – whether she had her own calf in the three years before the grandoffspring was born did not modify the beneficial effects she had. And the total number of calves she had in her lifetime was actually associated with reduced mortality in her grandcalf, with more experienced grandmothers (there were some with seven calves) having the biggest effect. From an evolutionary point of view, grandmothers increased their own fitness, or number of descendants, by enhancing the reproductive rates of their daughters and the survival of grandcalves. It makes sense, but it also plays out in our own families all of the time. I watch my own mother fly across continents to hold, change, feed and play with her grandchildren, and that's when the science becomes life.

And this is how it is for an elephant too. All of that growing up and growing old is happening at the same time to different individuals, within the context of a group: a family structure for the females, something more complex and dispersed for the males. But it's no wonder I see my elephants in my grandparents, and the other way around too. They understood all of this decades before I did. Because being an elephant, like being human, is about life experience and what you do with it. Karen McComb undertook a lovely study on the idea of matriarchs, the oldest females in a group, as repositories of social knowledge. Such matriarchs are often grandmothers. My own grandmother would baulk at such a grand and academic-sounding title as 'matriarch'. But she is indeed keeper of an incredibly tasty macaroni cheese recipe (with opaque instructions like 'make it like a custard') and the stories about her own mother, my great-grandmother, an eccentric and tough Londoner. Grandstar is connected in her village, she knows people. She has six grandchildren and five great-grandchildren at the last count. She could call herself a matriarch if she cared to.

The idea with the elephants is that, with age and experience, they accumulate a lot of knowledge about other elephants. To investigate this, researchers in Amboseli carried out a series of playback experiments, playing the vocalisations of elephants that the study herds would be familiar with and those they did not know. Imagine you hear the voice of someone you know: you might react by looking out for them. If you heard an individual that you didn't know, maybe you'd be suspicious and want to protect your family. Elephants can respond in similar ways, with curiosity or with defensive postures, gathering the calves in the centre and facing outwards, ready for any possible approach from a stranger. It's like they're making their own elephant castle. The team found that herds with older matriarchs were more likely to bunch together defensively in response to elephants that were unfamiliar to them than herds led by younger elephants. Those led by younger females might waste time responding defensively to familiar elephants, or missing the chance to get in formation in response to an unfamiliar elephant call. In a separate study, they showed that younger matriarchs also responded less defensively to playbacks of lion calls than herds led by older matriarchs. The older matriarchs also listened longer to the recordings of male lions than of female lions, suggesting they were interested in it. They stopped what they were doing to be aware of it, something their younger counterparts didn't do. A lion can be a real threat and a predator to calves, particularly a male lion, which can take a calf down alone. I know elephants take in a lot of sensory information including olfactory cues in their environment, but it's telling that older matriarchs reacted to those lion roars. When it comes to real life, the knowledge of these matriarchs really could be vital to survival and the continuation of the whole family.

It's not just knowledge about other animals that matriarchs possess, it's also knowledge about the landscape. Elephants are renowned for their spatial memory, and it's not just a saying. One study on elephants in Etosha National Park, Namibia, illustrates just this. The park includes a salt pan, which forms one of those saline desert places, almost moonscapes. The deepest depression fills fleetingly with water in the wet season, attracting flamingos and pelicans. But beyond the salty place, the majority of the land is woodland savannah, sprinkled liberally with mopane, that splash of green in dry places, with leaves like butterflies, sometimes home to thick white and black mopane worms, which, cooked right and with your eyes closed, taste like crisps. The researchers showed that elephants have highly directional movement, almost always to perennial water sources. The fact that elephants reached the nearest water source to them 90 per cent of the time indicates they do have a sense of where the water is. Elephants could start directing their movement to a water-hole on average 4.59 kilometres from the water and up to 49.97 kilometres away. This really illustrates both knowledge of the landscape and distribution of water sources and the development of strategies to minimise distance travelled to water. In the dry season, the distance of the decision point to move to water increased, indicating that the elephants adapted their behaviour to the more sparse water distribution at that time of year.

Elephants don't just use their memory when it comes to water. They can smell it, and they dig for it, accessing water that to many species would just not be available. But the holes they leave behind are available to others, such as humans, for example. Elephants are fussy about their water. I have experience of animals being that way. My dog will often ignore water, even when it's freshly filled in her bowl, but will happily drink water out of my hands. I know the feeling too. My kidney condition makes me a prodigious water drinker. The quantity I am supposed to drink to keep my own hormones in check becomes tedious. I try squeezing in lemon, pretending sparkling water is champagne. I imagine my kidneys, coated in cysts like bubble wrap, but much less fun to pop. I down yet another glass of water. The offending organs float away on a torrent; bloated, blood-filled deleterious detritus. Elephants also have a thirst to quench. In dry areas, it can be exasperating when elephants routinely damage boreholes to get water, even when there are pools of artificial water available. A study in north-western Namibia found that over a period of two years, the water from elephant 'wells' and boreholes that elephants had dug had lower levels of coliform bacteria when compared to the nearest surface water. This could be an example of how they adapt culturally to the landscape they are faced with, something so familiar from the human experience. I nodded when I read that the elephant wells weren't any less saline than the surface water, though. I know how much elephants have a taste for salt.

In areas where salt isn't readily available in water or plants, elephants will visit salt licks, and create and maintain depressions to eat salt-rich soils. A study in Hwange National Park, Zimbabwe, found that females spent more time eating soil at these sites than did males, probably because gestation and lactation increase their sodium requirements. In areas where rainfall doesn't meet the requirements elephants have for salt, these licks or salt soil sites become essential locations in the landscape. Another place young elephants are taken, another place matriarchs, leaders, have to remember. The distribution of minerals, which are required in smaller quantities than water, but are nonetheless vital for life, is fascinating. A study of the mineral composition of water in bais, natural forest clearings, in central Africa suggests that there's a complex mosaic of minerals at each site, and that elephants might use many such sites. The authors found that water from holes dug by elephants had higher mean mineral concentrations than the surface water. This extended to manganese, iron, sodium, phosphorus, strontium and others. The variance in mineral quantities makes it difficult for conservation because the bais are not necessarily interchangeable. And of course, because elephants are social, bais are centres for elephant interactions, particularly in the forested habitat that makes maintaining large groups difficult. For elephants, the social layer and ecology are not necessarily distinguishable. A bai could be both a place to drink and to associate with elephants that might not be part of the core family group but could be part of the wider social network. I sometimes think of a bai as a pub. A social place where you meet familiar characters, not just your family members. You might even noisily greet each other and drink together.

Because elephants live in multi-generational groups, the social and ecological knowledge of older females can be passed along. All of that knowledge about how to be an elephant, how to live like an elephant, is transferred, handed down. So that when a matriarch dies, her daughter can then lead. I think because we can look elephants in the eye with a certain amount of recognition, this doesn't seem like a spectacular feat. We also know that many animals have the capacity for learning, in that they can learn socially. They see others avoiding or exploiting a certain food and do this themselves. Even when we get to the tricky concept of culture, which we biologists tend to define as a behaviour that isn't passed on biologically but is learned, perhaps we're willing to see this in elephants or chimpanzees, though we are more surprised when it crops up in corvids or rats. What I suspect, however, is that it is our being human that limits our understanding of everything these grandmothers do and all that young elephants learn. We can track movements, map resources, record their rumbles going deep into the ground, and correlate them with behaviour. But we are probably missing a considerable amount. A lot of studies of culture in animals look at tool use in foraging. When chimpanzees were discovered to fish for termites using sticks in Tanzania's Gombe chimpanzee population, it was groundbreaking. Humans, the handy ape, who love to make things to solve problems, were not alone as tool users. We know now there is huge diversity and a wealth of chimpanzee culture, from a specific kind of hand clasp they exhibit, from grooming each other to using rocks to smash nuts. Elephants might also have a lot of cultural complexity and variation that we don't know about yet: dialects, use of salt licks, eating certain foods. For example, while there's no biological reasons killer whales can't eat a range of prey, some groups specialise and either only eat fish, such as salmon, or only eat marine mammals, such as seals. These 'cultural' or behavioural units could be important for conservation, for example the elephants in Namibia that live in desert areas. We might not find evidence for this specialism in their DNA, but it's in their behaviour. And we might not find some of the behaviours we think are vital for us. Perhaps we won't see as much evidence for tool use in elephants as there is in chimpanzees, because elephants have a trunk and they are massive. With that kind of strength and dexterity, perhaps fashioning tools is just not necessary. And tools are not hands. Elephants might not use them in the same way we use our hands because they want their smelling organs freely available. I hope we continue to be aware that being human is not the only way to be a successful long-lived organism, even long-lived mammal, and we should not let our perspective block out all of the other possibilities. It's actually a challenge to get out of your own head and skin and try to approach the world the way an elephant might.

As with culture, I think it's easy for us to understand elephants getting old, because they do it slowly, on a similar timeframe to ourselves. But there's growing evidence that ageing is not just for big, long-lived mammals. Even small garden birds like great tits age, their risk of dying increasing as they get older, and their reproductive success declining. It's not that living 'in the wild' means that you're condemned to dying young, being cut down in your prime (or that any of us is guaranteed not to die young). One of the things I really wanted to know when I first got into studying elephants was: do elephants stop reproducing? Do they have a menopause? They seem like a really logical species to test this idea, because they do have multiple generations living alongside each other. There is evidence that grandmothers improve the survival of some grand-offspring and the reproductive rates of their daughters. So potentially there's a possibility that they could invest in this grandmothering behaviour rather than directly in their own continued reproduction. And perhaps this would make sense if they were facing increased mortality rates with age, being less likely to raise their own calves to independence, because we are talking about a gestation of up to 22 months, plus a couple of years of lactation.

Mirkka Lahdenperä, a post-doctoral research fellow at Finland's University of Turku, got to the heart of this issue by comparing mortality rates and reproductive rates in humans and elephants. She used a dataset that Virpi had compiled with her team of church records that stretched back to before the industrial age, before people had access to modern medication we have now, where fertility and mortality were 'natural'. That's not to say there were no socio-cultural influences. These people were Lutheran and tended to be monogamous, not to divorce, to wait until they had a certain financial status before marrying. They were settled, they farmed, and that dependence on crops and crop failures affected mortality. They had male primogeniture as a mode for inheriting land. But, broadly, the patterns approximate what humans experienced for centuries: high infant mortality, lower in adults, and increasing with age, the curve shaped like a bathtub. And the fertility, even more interesting, kicking in late, in the late teens, but then there are very short inter-birth intervals. Humans can have a baby every other year if they're living in one place and in a good enough nutritional state. But as fast as that fertility curve rises, and as high as the peak is, it drops just as steeply. By the age of 40 in many, and 45 in most human females, there are no more births. There are always individual outliers, and they're particularly interesting: the very late or very early reproducers, and those who had lots of twins or even triplets. But the broad pattern was that a lot of female humans were alive after they stopped reproducing, even in that time. In contrast, the elephant reproductive rates never reached the human heights, and inter-birth intervals were longer, which makes sense because of the long gestation. In the elephants, there was no blank space where the reproductive rate had dropped off completely, but there were still many females alive. The reproduction and mortality curves dropped together. That post-reproductive lifespan just didn't seem to be there. Essentially, there was no elephant menopause.

This was very striking for me because I thought perhaps elephants would have a more pronounced post-reproductive life. I saw the benefits of investing in grand-offspring rather than continuing to reproduce, and it's been used as an explanation for human menopause. It essentially gives a female the chance to improve her fitness without taking the very energetically expensive strategy of producing more offspring of her own when her mortality risk is increasing. Indeed, such care for grand-offspring might be necessary, but not sufficient, for post-reproductive life to evolve. There are a couple of important things to point out about the human–elephant comparison. Firstly, that we have much better data for humans, even pre-demographic transition humans, than we do for elephants, even elephants living now. This is because individually identifying elephants and following them through their lives has only been done at a few sites in the past 50 years or so. I don't know why we didn't invest in this type of research before then, but that's the way it is; humans have other things to do. Nonetheless, having sites such as Amboseli and Samburu, and multiple sites rather than just one, is vital to understanding differences between elephant species and their lives in different environments. The Myanmar data is a treasure trove, but it also has its specifics: the elephants are Asian elephants, they are working elephants, about half were born in captivity, like baby Hannah, and some never reproduce, which is unusual in wild elephants. All scientists love data. They love multiple sites and it's not just greed and curiosity; we have to be really careful about generalising results from one place to a whole species.

In terms of identifying and tracking the lives of wild savannah elephants, Amboseli is the oldest and best known of these sites. In Amboseli, the researchers have also investigated reproductive ageing in females. The cast of researchers includes the indomitable Phyllis Lee, a no-nonsense American I first encountered when I was an undergraduate student at Cambridge and she was teaching a course on the behavioural ecology of primates. She was the tough external examiner to my PhD thesis and has since been a valued advisor and excellent presenter and discussant. I don't know if she'd want to be regarded as a matriarch, but she is doubtless a leader. She and her colleagues found that Amboseli elephants lived for about 16 years after they completed 95 per cent of their fertility, but they also noted that reproduction did not cease entirely until they were over 65. So there isn't that menopause relatively early on as in humans, but there's evidence for senescence there. Phyllis is also careful to note that, while 16 years might seem a long time, it might not be particularly longer than in other species, it's just a matter of scale; elephants are big with slow lives, so everything about their lives takes longer, but if you controlled for their size, they might not be outliers the way humans are in terms of post-reproductive lifespan. Similarly in Myanmar, the presence of a mother reproducing simultaneously with her daughter was associated with higher rates of reproduction in the daughter. That co-residence, co-presence, seems to be of benefit to elephants.

Where does this leave us? A female elephant has reproductive ageing like us, but not necessarily a reproductive cessation. We appear to be better off looking to whales for a comparison in that respect, given that orcas and short-finned pilot whales have evidence for a definite end to reproduction, not a slow-down like elephants. But I don't think it necessarily takes away from ageing and growing old in a family being a useful lens in which to see the elephant in us, and to see us in an elephant. As humans, we can make assumptions in both directions. We can highlight our uniqueness, when every biologist knows every species is unique: that's the whole point of attempting to define them. We can also make assumptions about other animals based on what we do. We have menopause, so why wouldn't other long-lived mammals? We rely on our sight as our domi-nant sense, so why wouldn't they? It's that kind of thinking that can limit us and make our efforts to understand others more laborious. What do orcas and humans have in common that elephants don't? One suggestion is that of reproductive conflict across generations. That's basically when mothers and their daughters reproduce at the same time. In some animal societies, an older mother might lose out in that competition, and then helping as a grandmother might be a more evolutionarily successful strategy than continuing to reproduce herself. But elephants don't do that; they continue to reproduce alongside their daughters.

Dispersal strategies could play a role. In orca pods, males and females stay in the family group, so over time, the pod fills with the offspring and grandoffspring of older females. In contrast, for female elephants, their relatedness to the rest of the herd probably doesn't increase in the same way over time, because their male offspring leave the group. In species in which females disperse and go into groups where they have no relatives and increase their relatedness to the group over time, the benefits of switching to grandmothering might be more apparent. And maybe that's a way humans lived in our evolutionary past, although humans take a huge range of mating and habitation patterns now. Ultimately, it's another one of the balancing acts in animal life: are females more successful, in terms of producing more descendants, if they continue to reproduce late into life or not? Whichever way, it could spread through the population through those genes they passed on, rippling through generations, beyond those they lived with, and helped directly.

I suppose that's the story thinly veiled by discussing ageing, that grandmothers leave. We're talking about the slide into death. I came across many stories about an orca named Granny, first caught when she was already fully grown in 1967 and followed from the 1970s until scientists stopped sighting her and presumed she was dead in 2017. I was sad that I would never see Granny when I read the last part. I felt a little cheated. All the effort to read about the studies she was involved in, the process of ageing her (she could have been over 100 when she died), the fame of her pod. I wanted to catch a little bit of that. Somehow, I wanted that connection, although she lived in a very different world to me, a submarine world that I'll never really understand, sharing with us just glimpses of it, when her shiny black back with its saddle-shaped whitish grey patch broke the surface and the humans watching noted it, and the distinctive half-moon nick in her dorsal fin. She was known and unknown, like all animals we study. A character, a staple, a symbol, carrying with her the memories of many thousands of people we know as Granny. Like many of them, a citizen of the twentieth century, although she would not have been aware of it. Her life is easier to approach than a barrage of overwhelming global events. But of course, Granny is in some ways even more than I thought of her. The very fact that her pod continues means that Granny still swims the ocean.

That's how it is with elephants too. And it's not just the continuation in her descendants. For elephants, they do seem to play some kind of role in the lives of other elephants after they are dead. In that respect, they have an afterlife. Elephants often, but not always, show an interest in their dead; in their bodies, certainly. Karen McComb and her colleagues performed a compelling experiment in Amboseli. They were interested in whether elephants reacted more to the skulls of elephants than to other objects, such as the skulls of rhino or buffalo, or a more neutral object, like a piece of wood. And indeed they did. They spent time smelling the elephant skulls and touching them with their trunks, and more rarely touching them with their feet or manipulating them. What's striking is they showed a preference for the elephant skulls over the other species and didn't prefer the buffalo skulls over rhino or vice versa. The elephant skulls were the outliers. They showed the most interest of all in ivory compared to the other objects. They also touched it more with their feet than they did with the other objects. The researchers also tested whether the elephants preferred a skull of the matriarch from their own group versus an unrelated matriarch. They didn't pick up an effect, but they were limited by sample size. These results are incredible. The capacity for elephants to distinguish appears to be nuanced and complex. We need further investigation. I was particularly struck by their interest in tusks. They selected them even more often than the elephant skulls. Such an important body part, a tool that elephants grow themselves, something that is strong, that is exposed in living elephants, that is an indirect cause of death, that decays very slowly after it. I wonder what the elephants see in those tusks, smell in them, feel in them.

Scientific accounts of elephants exhibiting behaviour associated with the actual death of a conspecific are rare. It's because they aren't frequently observed and the few instances could easily be dismissed as anecdotal. Iain Douglas-Hamilton and other Samburu researchers did publish one valuable account. It includes the association rates between the individuals in the study and tracking data from the collars of the elephants present, giving a broader picture of what happened from a spatial perspective. But the most interesting parts for me were the observations they made. I think of the vehicles, of the team who identify the elephants with such apparent ease. Of Iain, who flies the tiny planes and cares immeasurably about elephants because he knows what it is to follow them for years, to see them live and die, to look at them for the elephants they are. He and his team detailed the reaction of female elephants to a dying and eventually dead matriarch from the First Ladies group, named after the wives of US presidents. Her name was Eleanor.

The researchers found her with a swollen trunk and abrasions, probably having already fallen. One of her tusks was broken. She fell again. Another female (a matriarch from the group the Virtues), named Grace, came over to her. Grace had secretion streaming from the temporal gland on her head, and her tail was lifted. We know these signs mean some kind of excitement, the good kind or the bad kind. Grace sniffed with her trunk and touched Eleanor with her trunk and foot. Then, with her tusks, she lifted Eleanor up. Eleanor was unsteady on her feet, and Grace pushed Eleanor, so she might walk. But she didn't. That's when she fell again, for the last time. Grace stayed by her even when her own family left. She vocalised, nudged and pushed Eleanor. But Eleanor would not get up again, and Samburu darkened as the night crept over them.

By 11 am the next day, Eleanor had breathed her last. In the days following Eleanor's death, Iain's team carefully noted all of the individuals that interacted with her body. Many of them were not her relatives. Grace and her family, the Virtues, came by again. Another female, Maui, from the Hawaiian Island group of elephants, also showed a strong reaction to the remains. She sniffed, touched the body, and then tasted the tip of her trunk. She nudged and stood over the body, and rocked. She spent over eight minutes with it. On the sixth day after Eleanor had died, after her corpse had been visited by scavengers including lion, hyena, jackal and vulture, females from other groups continued to show an interest in it. Sage (from the Spice Girls group) spent three minutes touching and sniffing the body, with the collared matriarch Rosemary nearby. Biologists know they can explain the inter-actions of relatives in terms of evolutionary fitness because the individuals involved share so many genes. But it's always more complicated to understand why unrelated individuals interact. We don't know why so many females and no males visited Eleanor's body. But we know it happened.

Eleanor's closest association in life was with Maya, thought to be her daughter. Maya had been sighted over a hundred times and in 91 per cent of those Eleanor had also been there. This was Eleanor's family, the lives she impacted most. When Eleanor fell, Maya and the rest of the family had gone down to the river a kilometre and a half away to drink, perhaps not realising the matriarch was down. But they didn't move far away. The tracking data from Maya's collar shows she was close to Eleanor the day after she died, getting as close as 10 metres from her. The herd moved off, only to return to her the next day. When she died, Eleanor left a calf, a female, of around five and a half months old. The calf nuzzled her mother's remains. She walked around, trying to suckle from other calves and then looping back to her mother, confused. After twelve minutes, the Biblical Towns family approached and pushed away Eleanor's group, except for her calf, which stayed by Eleanor's body. The matriarch of the Biblical Towns, Babylon, had a very low level of association with Eleanor when she was alive. But they took an interest in her remains, sniffing it. Sucking up and analysing information in a way I don't understand. Maya, the calf, and the other family members stayed just a few hundred metres away. By the night, Babylon's group had moved on and they didn't return to the carcass site for a year. Perhaps they had been flexing their dominance by moving the First Ladies away from Eleanor's body, claiming something important. Either way, it was not for long. Maya and her family, as well as Sage's group, spent a lot of time in the area. Eleanor's calf was sighted several times with Maya, and she attempted to suckle from females in the group, but no successful attempts were observed. The calf died within about three months of Eleanor's death. She was too young to survive without a mother.

Reading Iain's account, again, I have to do the tightrope act a biologist has to perform as they walk along the fine line where what we sense and believe comes up against what we can measure consistently and define. We have the huge disadvantage of being enough of an elephant ourselves to know there's something going on, but not enough of an elephant to really know what it means to them. Words like 'grief' and 'mourning' seem too human to apply, but we don't know exactly what words to use in their place. I do agree we should be careful with our language. But I also think we should publish narrative data and accounts of specific incidents, as was done with Eleanor's death. In fact, I was thrilled to read descriptions of Asian elephant behaviour in response to deceased elephants in India, which is also nuanced, complex, and highlights that this behaviour is not only observed in African savannah elephants. It's really a problem with our reporting on it, rather than it not existing. It's a sample size of one event, or just a few events on the surface, but there were so many associations, interactions that came along with it. It's only in this way that we will get more insight into the response of elephants. We shouldn't use the risk of anthropomorphising elephants and what they do to distance ourselves from elephants and other animals that seem to respond to their dead, including giraffe, dolphins and chimpanzees. Equally, the human is not the gold standard or pinnacle of all behaviour. We should be careful not to limit those animals to the human, to only doing things that we might do and not more, and be aware how much they are capable of.

Ageing is so much part of our lives and so hard because we know where it ends. I wondered if Maya and the elephants could deal with it better than I can. As I started to write this chapter, my grandmother received a huge hamper of ginger-bread from my boyfriend in Germany: gingerbread houses, iced biscuits, elaborate shapes. She was delighted and phoned me to tell me all about it from her bungalow, the successor to the big house with the elephant on the door. My grandfather chipped in from the background. He prefers to shout out comments rather than holding the phone. As I wrote more, I took photographs for her of Bavarian houses, mountains in the background, castles on hills and maypoles to send to her. But I was late. By then, over just a weekend, she slipped a little from me, separated from me not only by fields, trees, sea, but by the inevitability of ageing. The chapter that was so linked to my grandmother would probably be not just hard for me to finish, but I started to worry she might never read it. This chapter that I had written out of sequence, drawn to it, followed not only Win Oo's decline as I had imagined but the arc of my own family, which with its backbone of three generations of women was starting to disintegrate. And all that I wrote about family, of the legacy of grandmothers, how to measure mourning, of Eleanor and Maya and collective memory, floated on the page as my eyes swam with tears. How could I ever explain how important my grandmother was? How important Eleanor was to Maya? I played the video of my grandmother recounting her macaroni cheese recipe ('you make it like a custard') over and over. I looked in the mirror and tried to see her in it. In some lights, in some photographs, I saw traces of her more optimistic face behind mine. Other times, I couldn't see past my own puffy eyes.

CHAPTER 10

Death and ivory

My death
That's so related to me as a wink to the eye.
Mongane Wally Serote

It was searingly hot and I absent-mindedly scratched the bright pink dots of itchy heat on my arm. I could smell the sticky sweet after-sun lotion that I swore I'd never buy again. It had sparkles in it, which caught the light and emphasised my heat rash. Below that top scent, there were middle notes of the honey from breakfast and base notes of earth, where the dry soil hit the river, that deep smell. I wonder if the elephants smelled all of that, processed it like me. They probably smelled much more: textures and signals and warnings I could never hope to capture. But it was sounds that disturbed my olfactory wanderings. The phone rang, and then the radio buzzed irritably. It was Iain's birthday in Samburu, 2010. But after those calls, no one was enjoying the lovingly assembled lunch outside, in the dining area that overlooked the river and the small hill the baboons paraded past. It was tense in the camp again. Iain looked a little drawn and wanted to take out the car to see what was going on. The news didn't seem good. Another elephant, Khadijah, had been shot. I grabbed my bag and camera. 'Can I come, please?' I clenched my jaw shut and set out. We drove along the bends of the river in silence until we found Khadijah's group. The elephants were clustered together and sheltered by trees. They weren't resting, but didn't seem particularly anxious. Their ears flapped, and the calves wandered between their mother's legs. Iain spotted Khadijah huddled in the group. He pointed out the bloodstains on her grey skin towards one of her back legs. If you didn't look too hard, perhaps you could believe she'd been splashed by water. She hadn't. She had been shot. But the good news was that she was up and about, she was with her family. We'd keep an eye on her.

I left Kenya before many of the elephants were killed, including ones named Resilience and Hope, mocking the fact they were named after virtues. I left before Khadijah recovered from her injuries and many did not; before there were so many orphans. About 20,000 African elephants a year were killed during the time I did my PhD and my first post-doctorate job. As I took on my own students and became a professor, the killing slowed, but continued. It seemed wrong for things to be going well for me as they became worse for the elephants. I only got anywhere because of them, and they were struggling. It wasn't the end of life that I wanted to study and understand.

I couldn't dismiss it as 'natural', even when I knew nature could be hard on individuals. It was preventable. It was caused by humans.

I knew the impact of the illegal killing, which we sometimes call poaching, had been enormous for the people working with elephants on the ground. They had to deal with their remains and the prospect of losing more. But those people did not give up, and over time, there was lessening, improvement, even some signs of recovery. By 2017, Kenya had seen a 40 per cent drop in the rate of illegal killing of elephants. Things were looking up on the 'demand' end of the ivory market too. China and the USA, the two biggest markets for ivory, had made an agreement virtually closing their legal ivory trade. Hong Kong was on the brink of ending the domestic ivory trade. At that time, I was in Hong Kong, feeling the changes. The city, like all cities, is in a perpetual liminal phase, constantly being reworked, redefined and reimagined. As well as all the solid physical components like the Peak Tram and the high-rise blocks, it also exists in our constantly shifting memory, perception and plans, and in us. Hong Kong is one of those onion places. Sometimes it's all glitzy towers, with glassy reflections of the harbour and rooftop bars, but there are layers there. The tea shop with drawers full of precious, delicate leaves in an old commercial building; the side street where you can buy all the dried fish you never knew you needed; the smattering of scars on one of the lions guarding the HSBC building, acquired from shelling during the Japanese invasion in

December 1941. The brash exterior is just a mask, and a transient one at that.

In among the layers that sometimes make you cry and, like the dried fish, are sometimes so tasty, there's a street called Hollywood Road. It's named after the plants, rather than the California film-making hub, but it has its own glamour. When I walk along Hollywood Road, it's a version of Hong Kong I love. The Man Mo Temple, green-roofed and red and candle-lit inside, dedicated to the twin gods worshipped by ambitious students seeking to pass Ming and Qing dynasty civil service exams, and frankly who all scholars should appeal to: the god of literature (his full name is Man Tai) and the god of war (Mo Tai). Forget pen and sword dichotomies: they are one. Inside, I lit incense and oil lamps. Every time I do that, I remember the Cantonese expression 'ga yau' and its direct Hong Kong English expression, 'add oil'. It's often described as denoting support and encouragement, which is correct, but of course this misses the physicality and the pungency of adding oil to burn. I will always use the original when I speak of Hong Kong. Outside, the humidity hit me again. The low-rise buildings around the temple spoke to the past, something Hong Kong seems to run from, but can't deny. I imagined them in a different time, before so much land was reclaimed, and the street was close to the shore. When it was bustling with traders, sellers, buyers, travellers, connecting China to the trade routes around the world. And the street was still buzzing. As well as the name on the street signs, you heard people refer to it as 'Antique

Street'. Shop after shop was crammed with beautifully crafted and elegant antiques, and I couldn't help noticing that many, many of them were made of ivory.

Everything in those shops I walked by was, or was said to be, documented ivory: legal, much of it predating the ban on internationally traded product, some of it is even from mammoths dug up from the Siberian tundra of my ice age dreams. We get the English word for elephant from the Greek *elephas,* which originally referred to ivory alone. And here it had come full circle, reducing the elephants to the shiny white teeth. I pressed my hand against the shop window, cold glass separating me from the ornaments and elephant ghosts. It was all heartbreakingly beautiful, elaborately carved. Sometimes I spotted a huge tusk, far too big to have been from a younger female like Khadijah, transformed into a dozen elephants walking along an enchanted road, all cut out of a death. It was hard to digest. Standing in front of one of those windows, I could hear the radio calling and see Khadijah's blood, Iain's furrowed brow, and the dead elephant with no face. Those images might not be directly linked to the antique shops at all. There are many people and many stories between an elephant walking in Africa and an antique shop in Hong Kong. But there is no denying that we, we humans, have used ivory as a material for hundreds of years and it all ultimately started with a dead elephant or even a mammoth. We have used it to make musical instrument parts, name stamps, religious icons, jewellery and many other objects. Humans love to make things, with those hands. Hands of opportunity and technology and hands we touch each other with. Hands that are sometimes stained with blood.

Even in areas associated with the ivory trade, including China and Hong Kong, most people don't own ivory, and many of them don't want to own it. The dominant perspectives historically were in favour of the ivory trade, often held by craftspeople, traders, collectors and those directly involved with ivory, and one against illegal trade, often taking a hard line in terms of law enforcement, but in favour of regulated legal trade. The latter was the position of government. However, in the run-up to 2017 and continuing now, an emerging 'anti-all trade' perspective emerged. This was encapsulated for me in the story of Belen Woo Fung, then aged eleven, addressing the Legislative Council of Hong Kong. She spoke clearly and confidently to legislators and officials, giving a persuasive speech on how elephants think and feel, describing their experiences of 'joy, anger, fear and depression'. While I was walking on Hollywood Road, she implored the members to support the 2017 ban legislation. She also opposed giving any compensation to ivory traders for their stocks, stating 'that would mean that we are rewarding the ivory traders and speculators for encouraging African poachers to illegally kill these beautiful creatures'.

I dipped into the subway at Sheung Wan. Looking at my phone, as everyone else on the MTR train was looking at theirs, I watched references to the proposed ban pop up on my social media feeds, some of them stating that human behaviour could change, and that elephant populations had disappeared in the past. They gave an example of the Roman empire's thirst for elephants leading to the extermination of populations in north Africa. One of many questions about elephants I think about and don't have a conclusive answer to, is where the Carthaginian general Hannibal's war elephants came from. Where were they from, the elephants that were famously marched across the Alps, or at least died somewhere along the way. Perhaps they were some of those now-vanished north African elephants or Asian ones from populations that have since disappeared. For example, one was named Surus ('the Syrian'). Descriptions of him suggest he was an Asian elephant with a single tusk. It's possible that he descended from the Asian elephants seized by the Ptolemies of Egypt during their campaigns in Syria. Hannibal, like so many others who used war elephants, probably knew they would incite fear, that their trumpets would be impressive, that soldiers could sit intimidatingly high, elevated on Surus above the battle. But elephants are, as George Orwell knew, expensive bits of machinery: the huge quantities of food and sheer bulk of elephants mean keeping them is no easy task. When the 37 war elephants were marched across the Alps in the year 218, most of them died in the harsh conditions. In fact, Surus was the last survivor. I think of him sometimes, standing alone and scarred, miles from the historical ranges of elephants, a stranger in a world of humans. An elephant in a different world, just as alien as a carved tusk in a shop on a sweltering day in Hong Kong. He lives on in stories, of course, with the elephants of childhood storybooks, myths, legends and half-recalled news articles, but he was more than them. He was real. I wondered how real elephants like Surus could live up to their own mythology, especially in places where they didn't live. But I needn't have worried. Posters and images of real elephants were at the centre of the campaigns to end the ivory trade.

The recent ivory trade, the post-colonial one, first spiked in the late 1970s and the early 1980s, fuelled by demand in Japan, North America and Europe. The changing legislation regarding the endangered status of elephants and trade in their parts meant it decreased in the 1990s. However, after the century turned, both seizures of ivory and killing of elephants picked up again, peaking around 2011. This twisting and turning of the trade in ivory was due to a perfect storm of factors: increasing demand, often from markets in Asia, boosted the price of ivory and other wildlife parts, like rhino horn. The prices spiked and peaked in 2014 with wholesale prices for raw ivory at US$2,100 per kilogramme. The increasing prices meant more people willing to participate in the trade, in the killing. There was enough instability or corruption to let it happen, even encourage it. The trade in illegal wildlife products, including ivory, was worth an estimated US$19 billion by 2018. And to some of us it just came crashing into our lives, like it came to me in my Samburu innocence. Other people had seen the storm whipped up before and were able to act fast in their attempts to stop it.

ELEPHANTS

The illegal killing of elephants wasn't evenly distributed across Africa and Asia. It never has been.

There have been instances in Myanmar, the much-loved place where I studied for my PhD. In the first nine months of 2018, fifteen wild elephants were killed. When scientists from the Smithsonian put tracking collars on nineteen elephants as part of their studies, they were shocked to find seven were killed within a year. And in a harrowing twist, it was not just their tusks, their teeth, that were removed. Some elephants killed in Myanmar had their trunks, skin, feet and ears cut off. In one disturbing photograph taken by the scientist and veterinarian Zaw Min Oo, an elephant had been skinned, possibly to make sanguineous jewellery from its skin. I forced myself to look at the photograph; the carcass was like a bloody white pupa. No ears, no face, it was de-elephantised. Dehumanised is a word. I don't think de-elephantised is one, but that's all I could think of to describe it. I couldn't smell that carcass, but I could imagine the stench, the thick rot in the dense Myanmar forest. It isn't how elephants are meant to die. It looked brutal and painful. I wondered where that elephant's family were now, what scars they bore. I understand the rush of emotion, the anger, the hopelessness a photograph like that evokes. I'm also conscious that the animals deserve better than my indignation. They deserve my action, and it's the cool heads that do best with that. Zaw and the Smithsonian team are still in the area, working with solutions in mind, tracking elephants. And they know this is serious. The paper they wrote describes it as a 'new crisis', and they are not fond of hyperbole. We have long thought illegal killing was just about ivory and much more common in Africa. But in Myanmar, when skin had also become the product, females, calves and family groups as well as males were targeted. This means elephants were facing the twin threats of habitat loss, which we associate more with Asia, and killing for their body parts. There are only 39,500 to 43,500 wild Asian elephants remaining at all. Surely we can spare making ourselves gory jewels from the few that are left.

Elephants in Africa also face highly variable risks depending on their location. The worst of the illegal killing in the most recent crisis was in central Africa, home of the forest elephants in the Democratic Republic of Congo and surrounding countries, and it radiated out to east Africa, cutting through savannah populations in Kenya and Tanzania. A report by the International Union for Conservation of Nature and Natural Resources (IUCN) in 2016, based on 275 estimates from across the African continent, put Africa's total elephant population at around 415,000, representing a decline of around 111,000, or 27 per cent, in ten years. The great elephant census covering 15 countries, published the same year, indicated a decrease of 144,000 elephants over a seven-year period, a rate of 8 per cent a year. The survey recorded the highest drops in numbers in Angola, Mozambique and Tanzania. In terms of absolute numbers, it highlighted surprisingly low numbers found in north-eastern Democratic Republic of Congo, northern Cameroon and south-west Zambia. The report emphasised that in those areas elephants could plausibly become locally extinct.

Where governance is weak, where instability is rife, elephants don't always face a good future. If things are in a state of disarray for humans, how can we expect wildlife to thrive? Where areas become safer and more stable for humans, elephant populations can recover. Make no mistake, the elephants can be part of our lives, even when we don't think they are. Although conservation might not be a top priority when it comes to tensions, war, regime changes and other political events, biodiversity experiences the consequences of all of these. It's not possible to escape it. I also don't want to paint a picture of homogeneous gloom and death. There are populations that the elephant census reported to be stable in South Africa, Uganda, parts of Malawi and a complex of protected areas covering parts of Benin, Niger and Burkina Faso. The last one, the largest population of elephants in West Africa, may have even been increasing. And things may have improved even further up to today.

Zakouma National Park in Chad is a great example. The killing started early there. In 2002, the park had around 4,000 elephants. Eight years later, there were just 400 left, a population so depleted it was estimated to be only two to three years from local extinction. At the height of the killing, members of the Janjaweed militia, mainly from Sudan, would go to Zakouma on horseback and camp out there for a couple of weeks, killing all the elephants they could find. Then they carried the tusks away in camel caravans over the border to

Sudan, credit for more violence. Some of the raw ivory was processed within the region and emerged as trinkets in markets in Khartoum. Something so big and impressive transformed into figures, bracelets, chopsticks, things for tourists to buy. Other reports suggested that some of the ivory could have been trafficked through other African countries, changing hands, with people transporting it away to Asia. It did not look hopeful for the elephants. But things did change. The people running Zakouma implemented policies such as overhauling staff, training of rangers, changing the ranger perspective that the Sudanese people on horseback killing elephants had supernatural powers, and ensuring the park was staffed all year round, even in the wet season. These things made a difference. In 2006, 900 fewer elephants than 2005 were counted; presumably many were lost to illegal killing. In 2011, only seven elephants were killed by humans, and through 2016 up to February 2017, there were no reported unlawful killings of elephants. These changes weren't easy, and six rangers lost their lives in retaliatory killings after they seized the weapons and ammunition of a group that are presumed to have killed four elephants. But they are significant changes. Calves have been born, first 50 in 2014 and 2015, then 70 in 2016. As much as they mark the end to the killing, the calves also herald a turning point and a recovery. Things can improve from a seemingly terrible situation, from so much blood spilt.

As with many things concerning elephants, the ivory trade is much bigger and even more complicated than the puzzling and gigantic elephants themselves. That's why I avoided this chapter for as long as I could and put it after growing old. Ivory isn't an elephant anymore. It's part of an elephant that tells us a lot more about ourselves than it does about the elephant. It's a process as well as an object, the process by which we commodify the elephant. Sam Wasser, a professor at the University of Washington, knows more than many scientists about this. He started off as an ecologist and has had a hugely varied career (brilliantly typified by using trained dogs to pick up the scent of killer whale faeces off Vancouver Island, and finding a way to collect the samples). One fantastic thing he does is tracing the origins of ivory using genetics. When I first went to Samburu, this was all new. I remember sitting in the elegant riverside camp of Elephant Watch, surrounded by colourful cushions and with Iain looking at me intently after reading the paper Sam led. Was this actually possible? Can one extract DNA from tusks? I told him I absolutely couldn't see why it shouldn't be possible. Tusks are teeth: process them correctly, grind them up without getting them too hot and denaturing (disrupting the structure of) the DNA, and it should work. Linking this to where they came from is slightly more complicated, but essentially requires having a library of genetic data from across the elephant range. If we think of it like a map, the DNA of elephants follows geographical clusters and clines, with elephants most similar genetically to the others they live alongside. It's the same kind of idea you get behind those kits in which you provide a human DNA sample, usually a cheek swab, and then a company extracts and analyses the DNA and gives you a little report on the geographic distribution of your ancestors. I did one myself, hoping for some surprises, and got very few, but I suppose that's the point. DNA can be a reliable predictor of an individual's origins and therefore isn't going to tell you anything new if you already know where you come from.

Sam and his team have made some incredible findings. I remember sitting in his talks in Berlin in 2013, and then again in Cambridge in 2017, lapping it all up. He was able to do an amazing thing that scientists are sometimes able to do, to gain the trust and respect of people who aren't scientists at all and to use that to get samples, do analyses and come up with results that are vital to us all. He was able to obtain ivory from huge seizures, often made in ports in Asia, and traced back where those tusks came from. One of his key findings was that the spate of killings that I witnessed in Kenya in 2010 was mobile. It started in central Africa and then moved east and south, a trail of the dead left behind it. Sam also has a keen eye and he noted that tusks usually come in pairs, but some big seizures only had one tusk of a pair, and the second might turn up in another raid, months later. He also noted the bags the ivory was found in. In one case, it was bags of rice from Pakistan. These hints led him to think that it was a few organisations doing a lot of moving of ivory to various ports in Asia, and transporting them within Africa before they were shipped. Sometimes the overland transport took them far from their origins, further across the horizon than any wandering elephant would ever normally go.

The bags he spotted struck a chord when he realised that identical ones were used to transport other illegal products: drugs, arms, things that have no business being anywhere near the remains of an elephant. But products are what they become, because, again, this afterlife is nothing to do with elephants' 'mourning' or the importance of a leader like Eleanor, and everything to do with the human world. That world in which desperation, desire, fear and endless appetites end with flashing lights and dirty-looking tusks, marked with the lip line of an elephant and being cut out by humans in uniforms from those tatty rice bags miles from Africa. Those are the ones Sam managed to extract DNA from. There are also tusks that are not intercepted and end up being carved, being sold, taking on a whole new symbolic power to the people who own them.

I realised pretty early on that the trade in ivory was not just impacting numbers of elephants. It was affecting social behaviour, family structure, the pool of potential mates and even the physical characteristics of the elephants left behind. I knew that to show this, measuring elephants was important. With my work with Simon and Virpi, we had weighed and measured elephants and then calculated how measurements taken from photographs correlated with these real-life measurements. I always found it neat that having a clear photograph of the chest taken from the side means that you can estimate the weight of an Asian elephant. When I first returned to work in Africa in 2015, drawn back by the peak in illegal killing and wanting to contribute something, tusk size as well as those body measurements were on my mind. It all made sense: to an elephant those tusks show age, as they grow year on year like the rings on a tree. They are a tool, they can be used to uproot a tree and help them to eat the roots and rhizomes beneath, or they can be used to strip bark. They can be used in a fight, like the one that killed Soshangane. Tusks can also potentially give us information about the overall health and condition of an elephant; they can only grow long and thick and strong if an elephant has enough energy to invest in doing this.

I thought of one of the elephants we saw most, Sizwe. Of all the photographs Michelle and her Elephants Alive team took in the reserves adjacent to Kruger, Sizwe featured in the most. I could see why. He was easy to identify not just because he was one of the oldest and biggest elephants in the area, but because he had incredible tusks. They didn't jut out and forwards. Instead they swept long and low, curving elegantly at the end, and they were beautifully symmetrical. His tusks made him look like an artist's impression of an elephant, but he was absolutely real, there with us, and often with some younger followers. He proved how real he was by breaking one of those tusks clean off. I watched the stump grow back over time, slowly starting to recapture some of that fabled symmetry. Sizwe was a classic. He would never go out of style, with or without those tusks, but I did wonder how the break affected him. Because they're not just sending us messages with their tusks, in all likelihood, we are picking up messages that they are sending to other elephants. By displaying their tusks, elephants could signal their quality as a mate, and females might even plump for males with larger than average tusks. We scientists often think about it like the tail of a bird of paradise: it's essentially an ornament which might actually impede the animal in some ways. For example, the tail could affect the flight of a bird or make it visible to predators, and the tusks of a male elephant can be very heavy to lug around or even affect how they are able to eat. But if there is a pressure for large ornaments from that female mating preference, they do persist, usually teetering on the edge between being large enough to be attractive and not so energetically expensive or high in mortality risk that it's not worth it for the animal. I was so excited to understand the variation in tusk size and integrate it into all the other work I was doing: on social networks with Derek, on population genetics with Tess, on trying to understand those vocalisations.

On a collaring operation in South Africa, I remember touching the tusks of an elephant, one of the few times I could measure those of a wild elephant with my own hands. The elephant was on his side, trunk outstretched, with two little sticks keeping the airway open, his ears flopped down. He lay on his side, breathing slowly and deeply. The researchers and assistants worked quietly and quickly around him, taking blood samples from his ear, fitting the collar around his neck, deep in concentration as they trimmed off the excess of the thick band so it wouldn't get in his way when he ate or moved around. A few wiry hairs sprung from the elephant's face. Then there were the tusks. They were thick and strong and smooth, colder than I thought they would be. I wrapped a tape measure around them at the lip line. This is how tusks are measured: all of this activity and effort, to measure something with a tape measure. I snapped a pair of latex gloves on; perhaps a faeces sample would also be useful, to tell us about his diet, to compare his DNA to that of other elephants in the area. Maybe there were even markers linked to those tusks and how big they are encoded in the DNA. With the agreement of the veterinarian and other researchers, I lubricated one gloved hand and slipped it into his anus. Usually there would be a dung ball waiting there, but this time, nothing. I gave up, pulled out the hand. It was covered in a sheen of liquid faeces and many tiny wriggling white worms, less than 5 millimetres long. I took a photograph of my hand and scraped what I could into a tube for preservation. But I could not bring myself to eat food with that hand for days, the little translucent worms occupying my mind. I definitely couldn't pick up my apple. But I loved that the experience had filled me with both awe and a little bit of disgust, and, as always, the elephant humbled me. Data points are not always easy to collect, and sometimes, it leaves you with nightmares about those little worms.

Some people told me this effort to study tusks might not be exciting, that tusk size only tells us something about the age of an elephant. And indeed, we did find tusk size increased with age. We also found within age classes that it varies. It's exciting we can capture this just from a snapshot. No fancy equipment, no knowledge of the distance of the subject from the camera is required. What's more, I thought we'd need to compare the tusk size to the overall size of the elephant, perhaps using their shoulder height, but in so many photos, elephants aren't standing straight on, or grass covers their feet. But in fact, just measuring the width of the head from a photo is tightly correlated enough with the other measures of body size that we can use that instead. This was a huge relief and opportunity, because scientists often only take photographs of an elephant's face and ears, like a mugshot to identify them, and it turns out these photographs can be used to estimate body size, age and tusk size.

These seemed like vital pieces of the puzzle to understand elephants, how their lives make sense. Behind it, there was something undeniable and difficult. Tusk size might be such an important marker for the overall health of an elephant, but potentially the same trait, which had been beneficial to elephants, was precisely what they were killed for in some cases. Elephants like Sizwe might be more attractive to people wanting to kill them for their tusks. Perhaps Sizwe was even lucky to have that tusk break, as it meant he could possibly have a less risky life. From a scientific perspective, humans looking for trophies and tusks were putting a huge pressure on the population. Basically, we were removing the ones with big tusks and making it safer to be a small-tusked elephant, and more likely for those elephants to reproduce. This pressure might be in the exact opposite direction to the preferences of elephants, and also reduces the variation in the population. We all know Sweetie has tiny tusks that never grew beyond his lip line, but in male African elephants, this is very rare. Perhaps it could become more common, and with it there could be shifts in those other things I wanted to understand: how elephants relate to each other, how their populations are structured from a genetic perspective and how they communicate with each other.

Studying tusk size in the face of the killing made perfect sense to me. African savannah and forest elephants usually grow tusks as adults, males and females alike, although the tusks of females are thinner and often shorter. In Asian elephants, females rarely have tusks reaching below their lip line, and tusklessness in adult males is more common. The way the tusk size of elephants is changing over time is best illustrated in populations that have experienced high levels of killing. One such example is Gorongosa in Mozambique. During the 15-year civil war, the elephant population was greatly reduced. From a peak of around 4,000, the population fell to the hundreds. What Joyce Poole, an eminent scientist and experienced researcher, and her team found was that the loss wasn't just about numbers of elephants. In elephants born since the war ended in 1992, a third of females had no tusks. In females who were aged over 25 years, ones that lived through the war, 51 per cent were tuskless. These numbers are striking, because we know that tusklessness only usually occurs in 2–4 per cent of female African elephants. Joyce noted that tusk size wasn't just an issue for males, but that in populations suffering high pressure from illegal killing, older females with their larger tusks start to be targeted and killed. This is a pattern often repeated in the range of African savannah and forest elephants. Famously in the Addo Park in South Africa, with elephant numbers increasing from a very small population after extreme pressure from hunting, the percentage of tuskless females reached 98 per cent of the 174 females in the early 2000s. The park now represents a pocket of elephants existing among vast stretches with no elephants at all. This kind of distribution, characterised by fragmentation of habitats and even isolation of populations, might become increasingly common if elephant numbers continue to decline. And I write 'continue to decline' as if there were no agents operating there. But it's not a forgone conclusion. There are still opportunities to make a difference.

Being completely tuskless, like Sweetie or some of the females in Gorongosa or Addo, is unusual for males. But reductions in tusk size over time have been observed in some elephant populations, and males could respond by having smaller tusks rather than none at all. For example, a study by Patrick Chiyo and his colleagues in Kenya compared tusk size of elephants before the severe pressure of elephants being killed for ivory in the late 1970s and early 1980s, to tusk size of those who survived and ones that were born during the population recovery in the 1990s. They carefully compared elephants from the same areas and also controlled for their body size, which we know is tightly correlated with tusk size. Overall, the tusk length in elephants declined by about 21 per cent in males and 27 per cent in females born during the recovery compared to the pre-killing population, and tusk length of survivors was 22 per cent shorter in males and 37 per cent shorter in females. This might be because elephants with shorter tusks were more likely to survive the spate of killing and produce calves, because they were the only ones left. So they then passed on their propensity for small tusks to the next generation. The circumference, or thickness, of tusks was also reduced by 5 per cent in male elephants born in the recovery, but not in females, and the tusk thickness of survivors was reduced by 8 per cent in males and 11 per cent in female survivors. I noted when I read the paper that in the elephants before the cull, many males had tusk lengths of 140 centimetres or more, whereas among the elephants born during the recovery, there were no tusks longer than 120 centimetres. I paused as I looked at the plots, and I knew it was sad. But I was also so proud of these researchers who were meticulously collecting and analysing data even when things were so hard.

It's important that things have changed, that we're not still in those days where a carcass could be found every week in Kenya by the Save the Elephants team. But it's also important not to be complacent about the changes that have been made. It would be utterly naïve to imagine that it could not happen again; the fact that it has not, as I write, is testament to a huge amount of effort and work by many people. Including people who live in areas the elephants live in. I don't like military metaphors in conservation, I very much think any battles being fought are usually with ourselves, so I won't say those people live on the frontline. They often live in incredibly beautiful places, they live in overlapping spaces with wildlife, or alongside it or interspersed with it. This can require a huge amount of effort, compromise and loss.

In Samburu, Shifra Goldenberg and her colleagues have been studying the response of the population to the enormous demographic shift it underwent after high rates of illegal killing. Almost every adult male elephant was killed, and about a third of the population in total. That leaves a lot of orphans behind. They found that orphan female elephants, those aged from six to seventeen years, who were not with their mothers, received more aggressive behaviour than non-orphan females. This could be as simple as being displaced from the spot they were standing in to being kicked or poked with tusks. Family groups are usually very cohesive and stable, but the killing of adult females affected this. Orphans were more likely than non-orphans to move into new groups, and when they did, they experienced more aggression than other female elephants their age. So recovery of numbers is just one part of the picture: the changing demography of elephants as a result of killing has trickle-down effects into social behaviour. It would be interesting to find out whether hormonally, in terms of the stress response, orphan elephants are also operating differently. Perhaps the effects on orphans will be passed down to the next generations, through their reproductive patterns, their ageing, behaviour or stress responses. I was talking to a friend in South Africa about these long-term, cross-generational impacts of killing, displacement, loss of family members. I was really concerned about the health of elephants in populations that had experienced so much upheaval. My friend wasn't a scientist, but he nodded with understanding. It's trauma, he said. The generations of fear, the experience of violence, the struggle to survive: he said he knew what they were feeling, that human South Africans had also struggled and felt those things.

Just over the border in Botswana, I didn't see the same scars, but I saw many elephants, from the watery Okavango Delta to the much more arid Makgadikgadi Pans National Park. Botswana has banned hunting and has been much more politically stable than some of the surrounding regions. There are around 130,000 elephants in the country, but for some people this is too much for the human population to handle. There's pressure to change. In Makgadikgadi we drove out and saw a group of huge males clustered around a waterhole. Their tusks weren't as impressive as some I had seen in Kenya or South Africa, perhaps due to the minerals available in the area. But all the elephants were tall, and they were in good shape. It was mainly an area for males, but when I arrived in 2016, some females had been seen. Perhaps this was a sign that despite its aridity, the area was moving from a peripheral elephant habitat, mainly occupied by young males, to one inhabited more widely. We drove on to the Boteti River. I got out of the car and walked in the fine bed of sand that made up the dry riverbed. There were footprints of big elephants there, perhaps of the males we'd just seen. Each was crackled with lines and crevices, some more oval, some more round.

My friends at the car shouted to me. What had started as a trickle became a steady flow of water. After more than a year of running dry, and against many predictions, the Boteti River was back. People hugged, squealed with delight. We stood in the forming pools, kicked up the water into hundreds of tiny crystals. We filmed the water trickling and then flowing past us in time-lapse, but it was happening so quickly we hardly needed it. Botswana at the time felt like a refuge, but even that country faces risks. In late 2018, there was concern when 87 fresh carcasses were found during a three-month census. There was disagreement over whether these represented 'poached' elephants, as suggested by representatives of the non-profits carrying out the surveys, or ones dying of natural causes, as the government proposed, who counted only 53 dead elephants. Just like the flowing and drying of the Boteti, its unpredictability makes me nervous as well as hopeful.

For some elephants, the ivory trade looms large in their lives, a grinning death with a wink of an eye. Others are less impacted and may be more likely to face a death at the hands of legal hunters or 'conflict' with humans such as being hit by a train running on a track that cuts across an elephant's habitual annual movement pattern, or killed by farmers or communities who see elephants as a risk to their crops, a risk to their lives. But it goes without saying that human activities and land-use choices have a huge impact on the number and distribution of elephants alive today. And we will continue to have that enormous ability to make a difference or to lose so much. For me, whether on Hollywood Road in Hong Kong, in Kruger or Kenya, I always feel the elephant right by me. And for an elephant, ivory is not ivory. It's a tusk, it's their tooth.

I was lying on the chair of my dentist, who just happens to be a childhood schoolfriend. It was the summer of 2018. In a break from my teaching, I decided to visit for a catch-up and dental check. She shone that dentist-bright light, gently prodded my gums and inspected my teeth. She announced with a smile that, despite me still having that baby canine tooth on the upper right side, all was in good shape, and I could keep up my cavity-free record. I felt rather smug. But for me, something that is personal and just about my health like my teeth, is a complex geopolitical issue when it comes to elephants. I think I am an elephant most of the time. But when there are government discussions about trade in my dentition, when it puts my life at risk to have these teeth, when the next generation's ability to eat their usual diet could be affected, then I might have much more of an idea what it's like to be one.

As with everything about elephants and us, it's hard not to trip over the glaring symbols and metaphors. I remember piles of ivory being stacked up in Nairobi to be set alight in front of the snapping world press. As the flames licked up, bright orange and a surprisingly dark grey smoke billowed up next to the impeccably dressed guards in khaki, I couldn't

245

help thinking what it really meant. On the one hand, it felt too close to burning books for me, all of that information within the tusks: not just the DNA that Sam teases out, but age, diet, layers of elephant life, the curves of each tusk like a fingerprint. But I understood a bigger statement needed to be made. That there had been so much killing, that it had to stop, that this product was not going to be allowed to be available. Even though the fire towered high, the ivory it consumed wasn't enough to fill demand, so the logical alternative seemed to be to make the supply cease to exist. Humans and their hands again. Those that we think we control fire with, but oftentimes end up burning ourselves, our places, the elephants. I wished I had a trunk instead.

CHAPTER 11

If it doesn't stay, we pay

The key to immortality is first living a life worth remembering.
Bruce Lee

I gave a talk in Berlin in the summer of 2018. It was typical of the kind of talk I do in that it was overstuffed and ambitious. Forget your camera, forget the frame, forget the gallery: I was aiming for the kind of big picture that makes the world look small – as if it's orbiting around an extraplanetary elephant, and we need to shoot you into space just to get an idea of it. After the talk, there were questions. This was normal, it wasn't just because of my exuberance. Some of them came in the form of statements. I was told, you clearly want elephants to exist. Do you think that affects your objectivity in your work? I was affronted. But it's not the first time I had been asked something similar. Does your position, your attachment, your emotion affect what you do? That's one of three types of questions not directly about elephants that I'm asked. The second one is, how do you work with people who think different things to you about elephants? Like people who kill them for fun or for ivory or because they think it's necessary for conservation. Thirdly, I am often asked, what do we need to do to ensure elephant species continue to survive or how can we be better at it? Often, especially in schools, I am asked very colourful questions, including whether elephants sneeze (yes), whether I've ever been charged by an elephant (yes, it was my fault for getting too close), and if I was scared by that (definitely yes). Yet the three questions about objectivity, about differing opinions and about how to ensure the continued existence of elephants are much harder to answer.

In Berlin, I thought about making a joke out of my response. I study elephant mortality; how subjective do I have to be for me not to be able to count how many elephants died? That's not a classification error, that's scientific fraud. I didn't say that. I said something vague about conveying a lack of objectivity not having been my intention. But I should have been clear. I did not particularly care about elephants until I realised they have extraordinary lives. I was interested in them the way you might be interested in anything with an unusual strategy. Like the first time you use a home-sharing service instead of a hotel or pay for something with your phone. It's different and kind of exciting because it's weird. But you get used to those things, and they become far less appealing. The elephants were the opposite. Now, I do care because I know too many elephants not to, and besides, you only have to know one to be invested. I think everyone who works in conservation cares or represents people who care. It might be that they were drawn to it because they are better people than me, people who saw sooner the importance of biological diversity to the functioning of the planet. They might have stumbled into it because they fell in love with an elephant called Yeager after breakfast and have never really been able to get over it, like me. They give their time and energy and work (often underpaid and undervalued) to understand the patterns and processes of nature and the ways in which it can continue. They care, that's okay. We care, that's okay. Talking about them caring, that's also okay. If no one cared, perhaps much less of the work would be done or even thought of.

I'm actually often more concerned about being too narrow in considering what scientists can contribute than about losing my 'objectivity', although these issues are entangled. If you refuse to acknowledge your bias, the personal, societal, structural context of your work, and get stuck in the trap of the scientist as the purely objective individual, separated from all the mess of the real world, discovering ultimate truths, you can also be limited. Certainly, that would be the case if you're studying threatened species. Scientists might avoid being seen to over-state the importance of their work, discussing its applications, discussing the urgency of it, because they don't want to call their scientific credibility, their objectivity, into question. It sounds too much like lobbying, too much like opinion.

Science is done by humans, and they can be motivated

by their interests, biases, world view, pet ideas, concerns about dying or saving their grandma or saving the world. I would say, that's all okay. As long as we're not cooking or picking or bending our data to fit our story, and we know when to remove ourselves because it's too close. Training to be a doctor because you watch a relative burdened by a disease is one thing; doing surgery on that same relative is another. I grew up in the world of anthropology, which meant when you read an ethnography written after 1990, you might well have to trawl through several pages of the author reflecting on their own biases and position before you get to what they actually observed. It frustrated me, it seemed indulgently narcissistic and unnecessary. But now I see the value in it as an exercise (even if not for its literary beauty).

It also helps me with the second group of questions I'm asked: those on how I navigate the different views of many people. Often, just saying it out loud helps me: I am biased because I'm not a mahout/oozie. I know a lot of oozies and mahouts, but I can never completely understand their relationships with their elephants, with their job. What I have observed and recalled could have been affected by the fact that I visited and caught glimpses of many places, but I grew up somewhere completely different. Some of the time, I am not with elephants at all. I'm by a warm fire at Pembroke College in Cambridge, thinking about that grey Henry Moore sculpture on the Foundress Court lawn, which from a certain angle looks like the back end of an elephant, and daydreaming about it coming to life. Pulling up the grass with its trunk, rumbling away to the porters, trumpeting in surprise at the tourists, eating the mulberry fruit from the top of the tree and entertaining the students much more than the poor college cat ever managed to.

I think it is a critical role of scientists to know they are one of many actors in a much bigger world. I always say a scientist in an ivory tower is not very useful, and a scientist studying elephants from an ivory tower is a joke. You have to know what's going on, even if it's hard to relate to. In Colorado, in the autumn of 2017, I had two magazines on my desk. They had beautiful photographs on the covers: on one, an elk backlit and bathed in orange light, antlers high-lighted against the sky like a proud tree. On the other, two mule deer, cuter and with those comically big ears, standing deep in the snow with their bifurcated antlers forking out: they were still grand, still evocative of an idealised wilderness landscape. It's a landscape we can be part of. I am not particularly outdoorsy, but the adverts inside for equipment, bags, tents and canteens, were interesting and spoke of prepa-ration, active participation, submersion, interaction. But I found it so alien. I didn't just struggle to relate because of my lack of camping expertise and because I haven't visited much of vast North America. It was because these magazines were for hunters.

At the start of it all, I had wanted to be a scientist to understand how I was human. I wanted to know what life and death were. This is so internal and essential and introspective. It was also too close for me to really look at it, in myself. It was only by seeing those things in elephants that I could remotely understand what it meant to be human. Facing mortality, having a family, getting old. It all made sense now because of elephants. They somehow even made me less afraid of my own mortality. In all likelihood, the disease I have will ultimately kill me. But that's fine. My death was a certainty from the start anyway. I just wouldn't or couldn't accept that until I studied elephants like Soshangane or Baby Carly's mother dying. And because they meant something, then perhaps I did too, even if just by association. I don't regret having to have this point of comparison with myself, because it's often easier to have a reference point. I didn't expect that I had to look up and out so far to see myself. But I ended up seeing elephants and humans around the world and realising I wasn't the only one who was human, they all were too, somehow. All of them were so vastly different and deeply familiar to me at the same time. This doesn't eliminate all the variation between individuals. I really think that variation in their lives, our lives, is what makes them so fascinating. And life certainly is not what I once thought it was: just a series of biological events between a beginning and an end. It is the story of each one of us.

All of this looking out, instead of looking at my own navel, meant I had to listen to a lot of people who were looking at the same elephants as me and seeing something else entirely. They saw gods, pests, hunting trophies, fantastic photographs, totems, money, threat, possibilities. I saw them see it. I questioned my own detachment, my methods, whether I even wanted to be human. I argued with them and myself. And I cared. I came to care so much about elephants. I couldn't help it. And for me that connection superseded the original curiosity. To the point that everything else I have done is nothing to me: diplomas, titles, words on a page on the internet, the many and various indexes created to calibrate my value as an academic. That is nothing. If there are no elephants, there are no people who see the elephant in each other. If the elephants are just parts that have been chipped into something very beautiful and very dead, then I don't want to be in that world.

Knowing all this about being human and being a scientist, we are required to tackle big questions that are sometimes related to significant issues. My favourite kind of question at the end of a talk like that one in Berlin is, what next? What's the next thing to do? In my mind, it's something like, how to coexist with elephants in this world? Sometimes, we can't even get to this topic because we distract ourselves with other things. I feel that we are still at the point in science when we always ask 'what?', what should we study, what's worth it, what species should we focus on? I don't believe in being prescriptive about these things. If the elephant helps you see the world and yourself, then study the elephant. Don't study, say, a frog because someone told you it's more important unless they really convinced you and now you think the frog holds the key to something (but absolutely don't kiss it on my account). Equally, if you're the person who is interested in how the slime mould, the Higgs boson, or an orchid contributes something, then do something, tell us, study, go for that. I don't think we should get trapped at 'what'. It's all interesting, it's all worth it. Because we have to get to the more difficult bit of what to do next: how do we do science on things that people really care about, or make people care about the science we do? How do we present it to people who might not care as much? How do we measure the outcomes when one of them is how we feel about the whole thing? How do we show we know we have a particular perspective, but we want to take into account lots of other ones? How do we deal with the fact that some essential participants in our studies might not be able to speak to us, or make themselves understood? How do we get the money to do all of this? How can we make sure it goes to the right places if it's really hard to determine what the most important things to do are and we can't do all the things we want to?

The short answer is, I don't know for sure. But I know that for me that the conservation and ecology of elephants is important. As to 'how', the easiest thing about how, is you can just say *together*, we do it *together*. And in more than one way. But that's really broad enough to be pretty much useless. What really helped me was teaching a summer school class on the topic at Pembroke College. I had twelve exceptional students who went beyond what I could possibly have imagined when we discussed the origins, the future, the perspectives in conservation. We discussed how it had always been a movement, a reaction, and it wasn't just a science, because the whole idea tumbles down if you don't start with there being some kind of value or benefit to having biodiversity around. You have to buy into this for the endeavour to make sense. And only then do the arguments about intrinsic or monetary value fall into place, or the discussion about how important the humans are in all of this. I have it easy with the role of humans, if I can say, well humans are elephants in many ways, elephants are human under many definitions of the word, so human-centred is animal-centred. But it's a serious point of debate and although I think we are agreed humans are the opportunity in conservation as well as creating many challenges, there are many nuanced and singular situations in which we constantly need to re-evaluate how the needs of humans and wildlife are met.

In terms of what to do about elephants, the good news is that so many people are doing something already. People are on the ground, like Iain Douglas-Hamilton in Kenya, working with elephant communities and populations. People are working on the wildlife trade, like Sam Wasser. Lots of people in big ivory markets like China, the Philippines, the United States do not want to buy ivory; in fact, they are actively pushing for markets to be closed. People are talking to communities who live near elephants, talking to them about their own plans and hopes for the future, listening to them and letting them take control of their own future with elephants for a change. I don't have to reinvent the wheel, these things are all how. It actually makes me happy to be around people who give so much. What is more difficult is a lingering perception that the whole thing is a luxury, that's why so many of us volunteer or choose to do this for a fraction of what could be earned in other sectors. And don't get me wrong, you can't put a price on what it feels like to drive out to the elephants and listen for them, catch that first glimpse of Bulumko by the road or Wild Spirit under a tree. But the issue with it is, only people who can afford to get access to that. And we are missing the enthusiasm, expertise, knowledge and experience of many people because we are choosing to treat biodiversity as a luxury and not a necessity.

I think this is one of our greatest mistakes. It would be a shame if all of our efforts in health, education, transport, agriculture and technology started to fail because our ecosystems no longer function. But this is a risk we are actively choosing to take every time we push back concerns about the environment. We have managed to generate a sense of urgency regarding climate change, and this is inextricably tied to changes in biodiversity. But we also need to carry on speaking about biodiversity and pushing for its prioritisation and for more investment in strategies that are already working. Donations are wonderful and generous and the support of the public buoys the mood of researchers when things are very tough. But we wouldn't want our healthcare and education systems run on voluntary donations alone. We need biodiversity to be right up there with them. For me, elephants will always be one of the most potent symbols of strength, cooperation, communication, and that power can be reduced to trinkets in our hands. So they work as my motivation and what I'm working for. But that they are a symbol is meaningful: no species, not even humans, can be considered in isolation. I recommend you choose your symbol or line or place and you use that in your consideration of biodiversity and conservation. Then use your own power to ensure it's a priority, in what you buy, how you vote, if you want to get involved yourself, any way you can.

When I reflect on what elephants mean to me, I know 'value' is really hard to measure if we're going to go beyond cold hard cash and our caring has to become part of the value. I heard the adage 'if it pays, it stays' concerning species conservation. Those species that brought some kind of immediate economic benefit would be maintained, and others allowed to fall by the wayside. Elephants might be a species with the potential to 'pay', based on their popular appeal or trophy hunters paying huge sums to kill them. But as a concept, the focus on financial value seemed perverse. When someone made the 'if it pays, it stays' argument directly to my face, I felt my temperature rise and my nails making crescents in my palms as I formed fists in my pockets. I fumed. Those that pay should be funding all the others. And anyway, perhaps the value, economic or otherwise, of the species the speaker was happy to lose, would only be felt after they disappeared. I replied, 'If it doesn't stay, we pay.' And I scowled. I don't think the statement had any effect, perhaps the scowl more so. But I stand by it.

In the same class in which the summer-school students

helped to chip some of the scales off my eyes, we talked a lot about the power of hope and optimism in conservation. I remember covering a slide in blue circles with frowns and dropping in one yellow circle with a smile. It's trite. But it actually means something, because I don't think we're at a point where we all have to be blue. I remembered a quote by that giant of ecology and nature writing, Aldo Leopold: 'One of the penalties of an ecological education is that one lives alone in a world of wounds.' I understand this, but I don't do it. I know some wounds can be healed, some scars can be beautiful, and that people alter landscapes. They always do. When it would be easy to slip into hoping for some kind of Malthusian disaster to solve our problems by rapidly cutting the human population, we have to go past that. We need to think, what can we do under these circumstances, valuing other people and respecting them?

I read out loud a headline to my class that read, 'This is the last generation that can save nature'. The accompanying report said global wildlife populations have fallen by 60 per cent in just over four decades. Accelerating pollution, deforest-ation, climate change, and other human-made factors have created a 'mind-blowing' crisis that wildlife is now enduring. It went on: current rates of species extinction are now up to 1,000 times higher than before human involvement in animal ecosystems became a factor. I often reflect that the opening to most conservation textbooks, even papers and reports, reads like the Fall of Humans and their Expulsion from Eden. Perhaps this makes sense; it's an image that many humans are familiar with. When I talk to my students about it, I often flash up on my screen a painting by Titian entitled *The Fall of Man*, which can be interpreted as our mistakes and knowledge, greed and poor decisions foretelling our ultimate apocalyptic demise. But perhaps this metaphor isn't always useful despite it being relatable to many, because it removes our agency and could play to our fears. It's so fatalistic that it could drive us to inertia in the face of something we see as unpreventable. It could be that stories of redemption, of change, of flux and freedom and choice might be more useful in a sphere in which we do have the ability to act. So this time, I didn't put up the Titian. Instead, I put up a picture of Yeager, facing the screen head-on. Despite all of the poaching in Samburu, and the loss of almost every male elephant of his age, he survived. Years after I saw him trunking the disused collars on one of my first days in the country, he still frequents the camp. He eats from the trees there, impresses another generation of interns. I said to the class, forget the framing for a moment and make your own frame for that headline. This is a huge opportunity and responsibility for us. This is a chance. Rather than falling helplessly, tragically, into the pattern of the Fall, we could do things differently. Because Yeager is still there. What a thing it is that you might be able to say: 'I was one of that last generation, and I did something.'

Epilogue

I always thought walking with elephants was a metaphor until I did it. Walking through the forest just outside of Chitwan National Park, Nepal, I saw footprints. Mine, booted, those of the elephant, Raj Kali, close behind me, sometimes obliterating my impressions with her padded ovals. And it wasn't just the ones we made; many others were present in the soft ground: the deep pads and claws of a tiger, the wild boar tracks, each like a pair of lungs, the various deer tracks. My favourites were the rhino ones, which I always think look like flowers, the bold kind you would use to decorate your camper van. Perhaps then you could drive it all the way to Nepal, the way Jim Edwards drove his car. The man who arrived in the 1960s and became co-owner of Tiger Tops. He then set about making it a lodge and a conservation landmark and must-go destination not just in Nepal, but in the world. His sons continue the Tiger Tops vision, building on it, adapting it, keeping Nepal and its wildlife at its very heart. I looked up at Raj Kali. She was tall with her double-domed forehead, and she was engaged primarily with eating. She had amber eyes that didn't rest on me. She was pulling whole branches down at a time and carrying them along as we walked. For me, this was thinking and brooding time; for her, it was a buffet.

I wanted to walk next to Raj Kali and the other elephants there because that's what I'd been trying to do all along, to explore a life lived in parallel. One with a knowing, a familiarity that might allow us to understand each other. But it wasn't just the two of us. Raj Kali's mahout was sitting up on her neck, in turns keeping her going and letting her eat, so in tune that I hardly even noticed half of what he was doing, as he communicated with Raj Kali through subtle foot movements. My guide, the astute naturalist Bhim Thanet, walked next to me, telling me about the medicinal qualities of plants, lending me his binoculars to see the peacocks, and leaning over to chat with me about what Raj Kali had for breakfast, poking a plant with his stick. And there was a forest around us, with all of those other inhabitants. Knowing about elephants is just like knowing anything else, more a process of realising how little one understands than feeling one is an authority, a master. Perhaps this doesn't matter much, though, because being around elephants makes you realise you have a responsibility to them, that it's worth having them in this world and that you can learn some things about them and marvel at all the rest that is unknown.

This felt like a comfortable end to the story, a rounding

off listening to the birds, to Raj Kali pulling off leaves and bark, to her mahout encouraging her onwards from time to time. Perhaps I'd watch her slightly deflated behind pad off into a hazy sunset, across the river. And I would muse about the villagers living there. The kids playing cricket and the women in bright clothes who lived right against the buffer zone where Chitwan National Park and the surrounding human-occupied areas meet, with its tigers and rhinos and all the other less glamorous inhabitants. How they could live peacefully there and how romantic it all was. But as it always is with elephants, there was more to it. We walked further. The forest transitioned to an area of tall grassland. That's where we saw the rhino.

I'm so accustomed to seeing animals like that from a vehicle, disguising myself as a big animal even though I'm just another skinny bipedal primate. It was a rush to be my own size, and a little scary. Bhim whispered that I should follow his directions. The rhino was a large male, looking as though he was armoured, a dull grey suit of the stuff. I half expected him to whip out a spiked mace. He looked as though he was a dinosaur, as though he wasn't meant for this time at all. His face was surprisingly cute, like a cartoon image of a rhino, but his bulk was real. Bhim manoeuvred us so we could see him and not disturb him. We stood between Raj Kali and a second female elephant, Dibya Kali. They were like giant sentinels and it was the safest place I could be.

The rhino sniffed and looked towards us, and I took two steps backwards into the foliage. Bhim turned to check on me because I was the only one who was behaving slightly erratically. I thought there would be strong tension, even aggression, but it wasn't there. There was respect in the air, a hint of adrenaline, sure, but the situation wasn't hectic. Not even my heartbeat was. The rhino was curious, and looked towards us, but didn't maintain the eye contact. Even when he turned away, I could barely get the right focus on my camera to take a picture. But then I took a breath and my picture. It was then that I really felt that human-elephant bond that I had talked about so much. With the mahouts on the elephants next to me, their elephants rumbling slightly and facing the rhino, I was able to see this rhino much closer than I ever could without a jeep or some other machine a human had made. It felt like cooperation between the humans and elephants. And I noticed that those lives in parallel, the reflections in the mirror, they really are one for so many. Raj Kali and her mahout can look in the mirror at the same time.

Being in an elephant camp is one of the best places to be. But while in Nepal, I was tempted outside. Beyond the grasslands where the mahouts cut and the elephants carried their own food back to camp. Beyond the silent cows, the vegetable garden, beyond the tiger footprints in the soft earth. Past the village dogs that followed us as far as they could and then stood wagging their tails as the Land Rover outpaced them. I went to Lumbini to see the birthplace of the Buddha. It seemed like a relaxing day out, but perhaps it wasn't, given the overtaking on the roads, particularly of the brightly painted trucks, whose drivers honked and then just seemed to go for it. We accelerated and braked our way, slightly hair-raisingly, to a tranquil place, where Bhim toured with me around the temples and shrines. We noted the architectural styles, the heaviness of some of the wooden doors and the window shutters. He declared with approval that the temples seemed expensive. We silently circumambulated the archaeological site surrounding the birthplace and paused by trees swathed in prayer flags. Hundreds of people sat with crossed legs and chanted. The prayer flags flapped, multi-coloured in the wind, and the chants of the pilgrims were carried on it. But for the gentle movement, the place had a stillness. It was calm. That very calmness after the car journey could have been jarring, but it was exactly what I needed to have a bit of clarity. I remembered to breathe.

On the way home, I saw a kitten dying on the road. It's tail twitched violently, and a bold splatter of blood streaked the road next to its black, furry soon-to-be remains. Sometimes, we are dying kittens on the road, with people walking past and going to the shop and setting down their motorbikes around us. Often, no one even sees it and silently repeats 'no' from behind the blurry window of a Land Rover the way I did that day. I can feel the desperation of fighting something so seemingly inevitable like the loss of biodiversity. But I don't think this has to be the way working in conservation has to feel; it's certainly not the way elephants or any other animals live. The circling kitten tail of desperation can be a whirl it's hard to get out of, but we have to. I choose to remember the other things I saw that day: the people chanting together, the temples from all around the world. The place where the Buddha was born, after, as the myth tells, his mother dreamed a white elephant had entered her, foretelling she would give birth to a great being.

I thought about walking with elephants. In Nepal, but in other places too, everywhere they should walk. Elephants walking with elephants in huge numbers, blurring the horizon, the way they haven't for decades. Because I live in hope for all elephants and their lives. I have hope for me and you. I see them in us every day, and I see us in them.

H.M.
Chitwan, Nepal, January 2019

Acknowledgements

I owe a great debt of gratitude to Dan Jones. Thank you for opening a door I didn't even know was there, Dan. I don't know what I did to deserve you believing in me, but I wrote this because of you. Pembroke College, Cambridge, also took a chance on me when I was a junior researcher and supported me through some of the most exciting and difficult times of my life. I met incredible people at Pembroke, shared food, stories and friendships. Pembroke gave me the freedom to engage in scholarship of a standard I never thought I could reach. The college and its staff, students and fellowship will always feel like home to me.

I thank everyone who has hosted me and made me a fortunate person who works with elephants. At Save the Elephants in Kenya, I thank Iain Douglas-Hamilton and Oria Douglas-Hamilton, David Daballen, Gilbert Sabinga, Chris Leadismo, Jerenimo Lepirei, Benjamin Loloju, Patrick Kabatha, Heather Gurd, Lembara and all the others who were with me in 2010. I thank Virpi Lummaa and all of her team, in the UK, Finland and Myanmar. Especially Carly Lynsdale, Aïda Nitsch, Jennie Crawley, Simon Chapman, Adam Hayward, Mirkka Lahdenperä, Khyne U Mar, Khin Than Win, Thuzar Thwin, Mumu Thein, all of the veterinarians, including Win Htut, Myo Nayzar, Aung Thura Soe, Htoo Htoo Aung, the mahouts, other staff and everyone who supported the University of Sheffield and University of Turku teams. I thank U Toke Gale, for writing about timber elephants.

In South Africa, I thank Michelle Henley, Director, Co-founder and Principal Researcher at Elephants Alive. Her study was started in 1996 with her mother, Cathy Greyling. I thank the rest of the Elephants Alive team, including Anka Bedetti de Kock, Christin Winter, Jessica Wilmot, Michelene Munro, Robin Cook and Ronny Makukule. I thank the APNR and landowners for access to the land. I thank Francesca Parrini and the Centre for African Ecology at the University of Witwatersrand for making me an honorary researcher. I thank Gilbert Pooley for hosting me, and for his advice on writing. I thank his friends, especially Busi, for helping me to rename some South African elephants in this text: it made me wish I had a hundred to name! I thank Cape Town, for being there when I had to get away from the field.

In Thailand, I thank John Roberts for letting me stay at the Golden Triangle and for his support of scientific work on elephants. I also thank everyone I worked with in Thailand, including all at Chiang Mai University for their help when I was getting to know Asian elephants. This includes Janine

Brown, Chatchote Thitaram, Nikorn Thongtip, Patcharapa Towiboon and many others. Joshua Plotnik has always offered to share his expertise, wry humour and rapport with me. I thank him and Cherry Keratimanochaya-Plotnik for the years of collaboration and friendship. I'm glad to be out in the world with you. In India, I thank Surendra Varma and his team for hosting me twice in Bangalore and letting me visit their site in Bannerghatta. I thank Friends of Elephants in Bangalore for allowing me to attend two inspiring sessions, and particularly for the thought-provoking presentation by Jyothy Karat. In Nepal, I thank Kristjan Edwards, Jack Edwards, Marie Stissing Jensen, Reshmi Parajuli, D.B. Chaudhary, Bhim Thanet, the mahouts and all the Tiger Tops team. I'm so glad I could walk with elephants with you.

I would like to thank Rachel Conway at Georgina Capel Associates for helping me to navigate the publishing process. I lived in four different countries during the few years it took to take this from a glimmer of an idea to a book. Rachel looked after me the whole way. Thank you for finding Myles Archibald and the excellent HarperCollins team for me. The UK-US Fulbright Commission allowed me to travel to and live in the USA and build my research links. In Colorado, I lived in the birdhouse in Fort Collins and loved being a member of George Wittemeyer's group. I will never forget the time I spent high up in the mountains there, especially with Aari Lotfipour and Christian Hermann. I also thank Louis Caron, Jeffrey Light and Christine Wong for hosting me in California. It was a dream and I love your cats. Louis

ACKNOWLEDGEMENTS

Caron, you are a beautiful man with a fine mind. I'm so happy I will always be your friend.

I also spent time at the Wissenschaftskolleg zu Berlin as a College for Life Sciences Fellow. I think it was one of the happiest times of my life. I will never forget the snow on the lakes and then the golden summer. Cambridge will always be my first love, but Wiko showed me you can have more than one great love. Thanks to Olivia Judson for her honest thoughts on my writing and to Alex Courtiol for his years of friendship. Thank you to the other fellows, the Wiko staff and to Berlin for providing me with a stimulating and friendly environment, wonderful facilities and room to think. I particularly recognise the support of Daniel Schönpflug in advising me on writing and publishing and Ulrike Pannasch for helping me with my career, life and everything else. Thanks to the Fishbeck Foundation for funding my workshop on applications of studying elephant behaviour and ecology, and thank you to all the attendees for the stimulating discussions. I also appreciate Wiko for allowing my dog Hershey to join us, which brought me boundless joy.

In Cambridge, I was generously supported by the Drapers' Company. I owe them great thanks for their interest in my research and the opportunities they afforded me. I was also fortunate to receive the Branco Weiss – Society in Science Fellowship, which funded my research, my travel, and my team for five years. I acknowledge the comradeship and inspiration of my co-fellows; I am lucky to be among them. The fellowship contributed to my understanding of male elephants, to the training of young scientists and to my ability to communicate my research. I am extremely grateful. I thank Bhaskar Vira, Andrew Balmford and the whole University of Cambridge Conservation Research Initiative and the Conservation Science and Conservation Evidence groups. Thank you for tolerating my frantic writing and for giving me some human contact during the times I mainly sequestered myself. Nick Davies was a great support to me in Cambridge, as well as an inspiration in everything from teaching to writing. Chris Smith was the most thoughtful and generous leader as Master while I was at Pembroke. His grace and effortless style made the college a beacon. I thank the 'young at heart' fellows for showing that we belong, and indeed all the Pembroke fellows for discussing ideas with me. Particular thanks goes to Alex Houen for suggesting I write an article for the *Martlet*, which I decided to make a love letter to elephant dung and somehow formed the beginning of this book. I thank those who supported me in the early days, especially Rob Foley and Lucio Vinicius. I needed to be an anthropologist to do this. In the Department of Zoology at the University of Cambridge, I thank Claire Spottiswoode, Rose Thorogood, Tim Weil and many others for all of their encouragement. I thank my Cambridge friends. Sertaç Sehlikoglu was the best neighbour I could have had, and I thank her for always feeding me, in every way. Sanne Cottaar for making me smile and always being uplifting. I thank Tae Hoon Kim for his charm and endless friendship. I thank Nikita Chiu for sticking at it and always believing in academia and in love, even when I couldn't handle either of them. Whenever I get one of your handwritten notes, I am filled with happiness.

At the University of Hong Kong, I thank the School of Biological Sciences, the Department of Geography and the Department of Politics and Public Administration. I thank David Dudgeon, for his advice and for always asking about my family – such an elephant thing to do. I thank Caroline Dingle for her spirit and her suggestions concerning the ivory chapter and all of my other colleagues for welcoming me. I thank Hong Kong for being the unique place it is. I could never capture you, but there is some part of you in my memory, in my heart and in my wildest dreams. I thank my team: Sylvia Lam, Even Leung, Derek Murphy, Sagarika Phalke, Annaëlle Surrealt, Hannah Tilley, Yifu Wang, Kaja Wierucka, Crystal Wong, Carolus Kwok (you count) and all the students and volunteers who have interacted with my laboratory. I especially thank my PhD student Teresa Santos, who read the full text of this book so thoughtfully and always acts with kindness. Thank you to my class at the Pembroke King's Summer School in 2018. You went beyond what I could have imagined and helped me to be a better writer. I want to give you all the greatest opportunities I possibly can, because you are all exceptional. I can't wait to see the spectacular things you do.

I have also been supported by the Leverhulme Trust, the University of Sheffield where I did my PhD, the Myanmar Timber Elephant Project, the Natural Environment Research Council, which funded my PhD studentship, the NRF-DST early career fellowship to South Africa for a post-doctoral fellowship, the Cambridge-Africa Alborada Trust, and the NERC Biomolecular Analysis facility. I thank my collaborators and colleagues involved in all of these funded projects.

I thank my family for never questioning my interests, but always finding a way to rib me about them. It's because of you that I can't be a boring dinner guest. My grandfather was a traveller long before I was, my father was a storyteller, my grandmother was an unrepentant optimist and my mother was a teacher. I'll never reach their standards, but I'm proud to be a bit of all of them. Having my siblings means I will never be alone and I think that's why I like social animals so much. I'm particularly thankful for my nieces and nephews. I wanted to write something that they might enjoy. Finally, I thank Matthias Egeler for his reading and re-reading of this text, for not letting me give up, and for his high tolerance of elephants and talk about elephants (and my off-key singing). I could not be an elephant without you. I love you. And I thank Hershey, of course, for being my shadow.

References

CHAPTER I

My first publication
Mumby, H.S., and Vinicius, L. (2008). Primate Growth in the Slow Lane: A Study of Inter-Species Variation in the Growth Constant A. *Evolutionary Biology* 35, 287–95.

CHAPTER 2

Aristotle on elephants
Peris, M. (2003). Aristotle's Notices on the Elephant. *Gajah* 22, 71–75.

CHAPTER 3

The Myanmar timber elephant database
Mumby, H.S., Courtiol, A., Mar, K.U. and Lummaa, V. (2013). Climatic Variation and Age-Specific Survival in Asian Elephants from Myanmar. *Ecology* 94, 1131–1141.
Mumby, H.S., Courtiol, A., Mar, K.U. and Lummaa, V. (2013). Birth Seasonality and Calf Mortality in a Large Population of Asian Elephants. *Ecology and Evolution* 3, 3794–3803.

CHAPTER 4

U Toke Gale's text on timber elephants
Toke Gale, U. (1974). *Burmese Timber Elephants*. Yangon, Burma: Trade Corporation.

The Quaternary extinction
Koch P.L., Barnosky A.D. (2006). Late Quaternary Extinctions: State of the Debate. *Annual Review of Ecology, Evolution, and Systematics* 37, 215–50.

Simon's master's degree research
Chapman, S.N., Mumby, H.S., Crawley, J.a.H., Mar, K.U., Htut, W., Thura Soe, A., Aung, H.H. and Lummaa, V. (2016). How Big Is It Really? Assessing the Efficacy of Indirect Estimates of Body Size in Asian Elephants. *PLOS ONE* 11, e0150533.
Mumby, H.S., Chapman, S.N., Crawley, J.a.H., Mar, K.U., Htut, W., Thura Soe, A., Aung, H.H. and Lummaa, V. (2015a). Distinguishing Between Determinate and Indeterminate Growth in a Long-Lived Mammal. *BMC Evolutionary Biology* 15, 214.

Jennie's master's degree research
Crawley, J.A.H., Mumby, H.S., Chapman, S.N., Lahdenperä, M., Mar, K.U., Htut, W., Thura Soe, A., Aung, H.H. and Lummaa, V. (2017). Is Bigger Better? The Relationship Between Size and Reproduction in Female Asian Elephants. *Journal of Evolutionary Biology* 30, 1836–45.

Measuring dung balls
Morrison, T.A., Chiyo, P.I., Moss, C.J. and Alberts, S. C. (2005). Measures of Dung Bolus Size for Known-Age African Elephants (*Loxodonta africana*): Implications for Age Estimation. *Journal of Zoology* 266, 89–94. doi:10.1017/S0952836905006631

Elephant teeth and using them to estimate age
Laws R.M. (1966). Age Criteria for the African Elephant. *East African Wildlife Journal*, 4 1–37.
Stansfield F.J, (2015). A Novel Objective Method of Estimating the Age of Mandibles from African Elephants (*Loxodonta africana*). *PLOS ONE* 10(5): e0124980. https://doi.org/10.1371/journal.pone.0124980

REFERENCES

Historic range of elephants
Sukumar, R. (2003). *The Living Elephants: Evolutionary Ecology, Behaviour and Conservation*. Oxford: Oxford University Press, 2003.

The great elephant census
Historic and current numbers
http://www.greatelephantcensus.com
Chase, M.J., Schlossberg, S., Griffin, C.R., Bouché, P.J.C., Djene, S.W., Elkan, P.W., Ferreira, S., Grossman, F., Kohi, E.M., Landen, K., Omondi, P., Peltier, A., Selier, S.a.J. and Sutcliffe, R. (2016). Continent-Wide Survey Reveals Massive Decline in African Savannah Elephants. *PeerJ* 4, e2354.

How much elephants eat
Shoshani, J., & Foley, C. (2000). Frequently Asked Questions About Elephants. *Elephant* 2(4), 78–87. Doi: 10.22237/elephant/ 1521732268

Jumbo
https://www.elephant.se/database2.php?elephant_id=2109
Shoshani, J., & Foley, C. (2000). Frequently Asked Questions About Elephants. *Elephant* 2(4), 78–87. Doi: 10.22237/elephant/ 1521732268

CHAPTER 5

Oestrus detection
Thitaram, C., Chansitthiwet, S., Pongsopawijit, P., Brown, J.L., Wongkalasin, W., Daram, P., Roongsri, R., Kalmapijit, A., Mahasawangkul, S., Rojanasthien, S., Colenbrander, B., Van Der Weijden, G.C. and Van Eerdenburg, F.J.C.M. (2009). Use of Genital Inspection and Female Urine Tests to Detect Oestrus in Captive Asian Elephants. *Animal Reproduction Science* 115, 267–78.

What is musth?
Poole, J.H. (1987). Rutting Behavior in African Elephants: The Phenomenon of Musth. *Behaviour* 102, 283–316.

Artifical insemination of an elephant
Brown, J. L., Göritz, F., Pratt-Hawkes, N., Hermes, R., Galloway, M., Graham, L. H., Gray, C., Walker, S. L., Gomez, A., Moreland, R., Murray, S., Schmitt, D. L., Howard, J., Lehnhardt, J., Beck, B., Bellem, A., Montali,

R. and Hildebrandt, T. B. (2004). Successful Artificial Insemination of an Asian Elephant at the National Zoological Park. *Zoo Biol.*, 23: 45–63. doi:10.1002/zoo.10116

Thongtip, N., Mahasawangkul, S., Thitaram, C., Pongsopavijitr, P., Kornkaewrat, K., Pinyopummin, A., Angkawanish, T., Jansittiwate, S., Rungsri, R., Boonprasert, K., Wongkalasin, W., Homkong, P., Dejchaisri, S., Wajjwalku, W. and Saikhun, K. (2009). Successful Artificial Insemination in the Asian Elephant (*Elephas maximus*) Using Chilled and Frozen-Thawed Semen. *Reproductive Biology and Endocrinology* 7.

Thongtip, N., Saikhun, J., Mahasawangkul, S., Kornkaewrat, K., Pongsopavijitr, P., Songsasen, N., and Pinyopummin, A. (2008). Potential factors affecting semen quality in the Asian elephant (Elephas maximus). *Reproductive Biology and Endocrinology* 6, 9.

Parasites

Lynsdale, C.L., Mumby, H.S., Hayward, A.D., Mar, K.U. and Lummaa, V. (2017). Parasite-Associated Mortality in a Long-Lived Mammal: Variation with Host Age, Sex, and Reproduction. *Ecology and Evolution* 7, 10904–15.

Lynsdale, C.L., Santos, D.J.F.D., Hayward, A.D., Mar, K.U., Htut, W., Aung, H.H., Soe, A.T. and Lummaa, V. (2015). A Standardised Faecal Collection Protocol for Intestinal Helminth Egg Counts in Asian Elephants, *Elephas maximus*. *International Journal for Parasitology: Parasites and Wildlife* 4, 307–15.

Asian elephant calves

Lahdenperä, M., Mar, K.U. and Lummaa, V. (2015). Short-Term and Delayed Effects of Mother Death on Calf Mortality in Asian Elephants. *Behavioral Ecology*.

Mar, K.U., Lahdenperä, M. and Lummaa, V. (2012). Causes and Correlates of Calf Mortality in Captive Asian Elephants (*Elephas maximus*). *PLOS ONE* 7, e32335.

Lahdenperä, M., Mar, K.U., Courtiol, A. and Lummaa, V. (2018). Differences in Age-Specific Mortality Between Wild-Caught and Captive-Born Asian Elephants. *Nature Communications* 9, 3023.

REFERENCES

CHAPTER 6

Male elephant bonds
Goldenberg, S.Z., De Silva, S., Rasmussen, H.B., Douglas-Hamilton, I. and Wittemyer, G. (2014). Controlling for Behavioural State Reveals Social Dynamics Among Male African Elephants, *Loxodonta africana*. *Animal Behaviour* 95, 111–19.

Genetic and social relatedness
Chiyo, P.I., Archie, E.A., Hollister-Smith, J.A., Lee, P.C., Poole, J.H., Moss, C.J. and Alberts, S.C. (2011). Association Patterns of African Elephants in All-Male Groups: The Role of Age and Genetic Relatedness. *Animal Behaviour* 81, 1093–99.

Derek's paper on social networks
Murphy, D., Mumby, H.S. and Henley, M.D. (2019). Age Differences in the Temporal Stability of a Male African Elephant (*Loxodonta africana*) Social Network. *Behavioral Ecology* arz152, https://doi.org/10.1093/beheco/arz152

Tess's paper on population genetics
Santos, T.L., Fernandes, C., Henley, M.D., Dawson, D.A., Mumby, H.S. (2019). Conservation Genetic Assessment of Savannah Elephants (*Loxodonta africana*) in the Greater Kruger Biosphere, South Africa. *Genes* 10, 779.

CHAPTER 7

Mirror self-recognition in elephants
Plotnik, J.M., De Waal, F.B.M. and Reiss, D. (2006). Self-Recognition in an Asian Elephant. *Proceedings of the National Academy of Sciences of the United States of America* 103, 17053–57.

Lateralisation of trunk movements
Haakonsson, J.E. and Semple, S. (2009). Lateralisation of Trunk Movements in Captive Asian Elephants (*Elephas maximus*). *Laterality: Asymmetries of Body, Brain and Cognition* 14, 413–22.

Chimpanzee self-recognition
Gallup, G.G. (1970). Chimpanzees: Self-Recognition. *Science* 167, 86.

ELEPHANTS

Eurasian magpie mirror self-recognition
de Waal, F.B.M. (2008). The Thief in the Mirror. *PLOS Biology* 6, e201.

Dolphin mirror self-recognition
Reiss, D. and Marino, L. (2001). Mirror Self-Recognition in the Bottlenose Dolphin: A Case of Cognitive Convergence. *Proceedings of the National Academy of Sciences* 98, 5937.

Cleaner wrasse mirror self-recognition
Kohda, M., Hotta, T., Takeyama, T., Awata, S., Tanaka, H., Asai, J.-Y., and Jordan, A.L. (2019). If a Fish Can Pass the Mark Test, What Are the Implications for Consciousness and Self-Awareness Testing in Animals? *PLOS Biology* 17, e3000021.

Cooperation in elephants
Plotnik, J.M., Lair, R., Suphachoksahakun, W. and De Waal, F.B.M. (2011). Elephants Know When They Need a Helping Trunk in a Cooperative Task. *Proceedings of the National Academy of Sciences* 108, 5116–5121.

Cooperation in chimapnzees
Crawford M.P. (1937) The Cooperative Solving of Problems by Young Chimpanzees. *Comparative Psychological Monographs* 14, 1–88.

Amboseli crop foraging
Chiyo, P.I., Lee, P.C., Moss, C.J., Archie, E.A., Hollister-Smith, J.A. and Alberts, S.C. (2011). No Risk, No Gain: Effects of Crop Raiding and Genetic Diversity on Body Size in Male Elephants. *Behavioral Ecology* 22(3), 552–58.

Asian elephant personality
Seltmann, M.W., Helle, S., Adams, M.J., Mar, K.U. and Lahdenperä, M. (2018). Evaluating the Personality Structure of Semi-Captive Asian Elephants Living in their Natural Habitat. *Royal Society Open Science* 5, 172026.
Seltmann, M.W., Helle, S., Htut, W. and Lahdenperä, M. (2019). Males Have More Aggressive and Less Sociable Personalities than Females in Semi-Captive Asian Elephants. *Scientific Reports* 9, 2668.

REFERENCES

Lee, P.C. and Moss C.J. (2012). Wild Female African Elephants (*Loxodonta africana*) Exhibit Personality Traits of Leadership and Social Integration. *Journal of Comparative Psychology* 126, 224–32. doi: 10.1037/a0026566

Big 5 personality model
Mccrae, R.R. and John, O.P. (1992). An Introduction to the Five-Factor Model and Its Applications. *Journal of Personality* 60, 175–215.

Empathy in elephants
Plotnik, J.M., and De Waal, F.B.M. (2014). Asian Elephants (Elephas maximus) Reassure Others in Distress. *PeerJ* 2, e278.

Dwarf elephants
Sen, S. (2017) A Review of the Pleistocene Dwarfed Elephants from the Aegean Islands, and their Paleogeographic Context. *Fossil Imprint* 73, 76–92.
Herridge, V.L. (2010) Dwarf Elephants on Mediterranean Islands: A Natural Experiment in Parallel Evolution. PhD thesis, University College London.

Olfaction
Niimura, Y., Matsui, A. and Touhara, K. (2014). Extreme Expansion of the Olfactory Receptor Gene Repertoire in African Elephants and Evolutionary Dynamics of Orthologous Gene Groups in 13 Placental Mammals. *Genome Research* 24, 1485–96.

Do elephants follow humans pointing?
Plotnik, J.M., Pokorny, J.J., Keratimanochaya, T., Webb, C., Beronja, H.F., Hennessy, A., Hill, J., Hill, V.J., Kiss, R., Maguire, C., Melville, B.L., Morrison, V.M.B., Seecoomar, D., Singer, B., Ukehaxhaj, J., Vlahakis, S.K., Ylli, D., Clayton, N.S., Roberts, J., Fure, E.L., Duchatelier, A.P. and Getz, D. (2013). Visual Cues Given by Humans Are Not Sufficient for Asian Elephants (*Elephas maximus*) to Find Hidden Food. *PLOS ONE* 8, e61174.

Olfaction paper with Josh
Plotnik, J.M., Brubaker, D.L., Dale, R., Tiller, L.N., Mumby, H.S. and Clayton, N.S. (2019). Elephants Have a Nose for Quantity. *Proceedings of the National Academy of Sciences*, 201818284.

ELEPHANTS

CHAPTER 8

Seasonal variation in stress

Mumby, H.S., Mar, K.U., Thitaram, C., Courtiol, A., Towiboon, P., Min-Oo, Z., Htut-Aung, Y., Brown, J.L. and Lummaa, V. (2015). Stress and Body Condition are Associated with Climate and Demography in Asian Elephants. *Conservation Physiology,* 3.

Elephant voices database

https://www.elephantvoices.org/elephant-communication/acoustic-communication.html

Elephant imitating human speech

Stoeger, Angela S., Mietchen, D., Oh, S., De Silva, S., Herbst, Christian T., Kwon, S. and Fitch, W.T. (2012). An Asian Elephant Imitates Human Speech. *Current Biology* 22, 2144–48.

Infra-sound

O'Connell-Rodwell, C.E., Hart, L.A and Arnason, B. (2001). Exploring the Potential Use of Seismic Waves as a Communication Channel by Elephants and Other Large Mammals. *American Zoologist* 41, 1157–70.

O'Connell-Rodwell, C.E., Arnason, B. and Hart, L.A. (2000). Seismic Properties of Asian Elephant (*Elephas maximus*) Vocalizations and Locomotion. *Journal of the Acoustical Society of America* 108, 3066–72.

O'Connell-Rodwell, C.E., Wood, J.D., Rodwell, T.C., Shriver, D., Puria, S., Partan, S.R., Keefe, R., Arnason, B.T. and Hart, L.A. (2006). Wild Elephant (*Loxodonta africana*) Breeding Herds Respond to Artificially Transmitted Seismic Stimuli. *Behavioral Ecology & Sociobiology* 59, 842–50.

O'Connell-Rodwell, C.E., Wood, J.D., Kinzley, C., Rodwell, T.C., Poole, J.H. and Puria, S. (2007). Wild African Elephants (*Loxodonta africana*) Discriminate Seismic Alarm Calls of Familiar Versus Unfamiliar Conspecifics. *Journal of the Acoustical Society of America* 122, 823–30.

https://news.stanford.edu/pr/01/elephants37.html

Bouley. D.M., Alarcon, C., Hildebrandt, T. and O'Connell-Rodwell, C.E. (2007). The Distribution, Density and Three Dimensional Histomorphology of Pacinian Corpuscles in the Asian Elephant (*Elephas maximus*) Foot and their Potential Role in Detecting Seismic Information. *Journal of Anatomy* 211, 428–35.

REFERENCES

O'Connell-Rodwell, C.E., Wood, J.D., Rodwell, T.C., Shriver, D., Puria, S., Partan, S.R., Keefe, R., Arnason, B.T. and Hart, L.A. (2006). Wild Elephant (*Loxodonta africana*) Breeding Herds Respond to Artificially Transmitted Seismic Stimuli. *Behavioral Ecology & Sociobiology* 59, 842–50.

Elephant sleep
Gravett, N., Bhagwandin, A., Sutcliffe, R., Landen, K., Chase, M.J., Lyamin, O.I., Siegel, J.M. and Manger, P.R. (2017). Inactivity/Sleep in Two Wild Free-Roaming African Elephant Matriarchs – Does Large Body Size Make Elephants the Shortest Mammalian Sleepers? *PLOS ONE* 12, e0171903.

Elephant listening project
http://elephantlisteningproject.org

Acoustic monitoring of elephants
Wrege, P. H., Rowland, E. D., Keen, S. and Shiu, Y. (2017), Acoustic Monitoring for Conservation in Tropical Forests: Examples From Forest Elephants. *Methods in Ecology and Evolution* 8, 1292–1301. doi:10.1111/2041-210X.12730

Rumbles
Stoeger, A.S. and Baotic, A. (2016). Information Content and Acoustic Structure of Male African Elephant Social Rumbles. *Scientific Reports* 6, 27585.

Elephants distinguishing between humans
Mccomb, K., Shannon, G., Sayialel, K.N. and Moss, C. (2014). Elephants Can Determine Ethnicity, Gender, and Age from Acoustic Cues in Human Voices. *Proceedings of the National Academy of Sciences* 111, 5433–38.

Crop foraging
Sukumar, R., and Gadgil, M. (1988). Male-Female Differences in Foraging on Crops by Asian Elephants. *Animal Behaviour* 36, 1233–35.

Humans killed in India by elephants
https://www.theguardian.com/environment/2017/aug/01/over-1000-people-killed-india-humans-wildlife-territories-meet

Living with the Wild *documentary on humans and wildlife, including the death of Muthaya*
By Jyothy Karat
https://www.jyothykarat.com/blog/2018/7/10/living-with-the-wild

CHAPTER 9

Pace of elephant life
Hayward, A.D., Lahdenperä, M. and Lummaa, V. (2014). Early Reproductive Investment, Senescence and Lifetime Reproductive Success in Female Asian Elephants. *Journal of Evolutionary Biology* 27, 772–83.

Grandmothers improving calf survival
Lahdenperä, M., Mar, K.U. and Lummaa V. (2016). Nearby Grandmother Enhances Calf Survival and Reproduction in Asian Elephants. *Scientific Reports* 6, 27213.
McComb, K., Moss, C., Durant, S.M., Baker, L. and Sayialel, S. (2001). Matriarchs as Repositories of Social Knowledge in African Elephants. *Science,* 292, 491–94.

Elephants reacting to lions
Mccomb, K., Shannon, G., Durant, S.M., Sayialel, K., Slotow, R., Poole, J. and Moss, C. (2011). Leadership in Elephants: The Adaptive Value of Age. *Proceedings of the Royal Society B: Biological Sciences* 278, 3270–76.

Salt pans
Polansky, L., Kilian, W. and Wittemyer, G. (2015). Elucidating the Significance of Spatial Memory on Movement Decisions by African Savannah Elephants Using State-Space Models. *Proceedings of the Royal Society B: Biological Sciences* 282, 20143042.

Elephants digging holes
Ramey, E.M., Ramey, R.R., Brown, L.M. and Kelley, S.T. (2013). Desert Dwelling African Elephants (*Loxodonta africana*) in Namibia Dig Wells to Purify Drinking Water. *Pachyderm* 53, 66–72.

REFERENCES

Elephants getting minerals from soil and water
Holdø, R.M., Dudley, J.P. and Mcdowell, L.R. (2002). Geophagy in the African Elephant in Relation to Availability of Dietary Sodium. *Journal of Mammalogy* 83, 652–64.
Metsio Sienne, J., Buchwald, R. and Wittemyer, G. (2014). Differentiation in Mineral Constituents in Elephant Selected Versus Unselected Water and Soil Resources at Central African Bais (forest clearings). *European Journal of Wildlife Research* 60, 377–82.

Culturally specific units in killer whales
Riesch, R., Barrett-Lennard, L.G., Ellis, G.M., Ford, J.K.B. and Deecke, V.B. (2012). Cultural Traditions and the Evolution of Reproductive Isolation: Ecological Speciation in Killer Whales? *Biological Journal of the Linnean Society* 106, 1–17.

Single population of desert elephants
Ishida, Y., Van Coeverden De Groot, P.J., Leggett, K.E.A., Putnam, A.S., Fox, V.E., Lai, J., Boag, P.T., Georgiadis, N.J. and Roca, A.L. (2016). Genetic Connectivity Across Marginal Habitats: The Elephants of the Namib Desert. *Ecology and Evolution* 6, 6189–6201.

Reproductive ageing of elephants
Lahdenperä, M., Mar, K.U. and Lummaa, V. (2014). Reproductive Cessation and Post-Reproductive Lifespan in Asian Elephants and Pre-Industrial Humans. *Frontiers in Zoology* 11, 54.
Lee, P.C., Fishlock, V., Webber, C.E. and Moss, C.J. (2016). The Reproductive Advantages of a Long Life: Longevity and Senescence in Wild Female African Elephants. *Behavioral Ecology and Sociobiology* 70, 337–45.

Ageing in orca pods
Brent, Lauren J.N., Franks, Daniel W., Foster, Emma A., Balcomb, Kenneth C., Cant, Michael A. and Croft, Darren P. (2015). Ecological Knowledge, Leadership, and the Evolution of Menopause in Killer Whales. *Current Biology* 25, 746–50.
Croft, D.P., Johnstone, R.A., Ellis, S., Nattrass, S., Franks, D.W., Brent, L.J.N., Mazzi, S., Balcomb, K.C., Ford, J.K.B. and Cant, M.A. (2017). Reproductive Conflict and the Evolution of Menopause in Killer Whales. *Current Biology* 27, 298–304.

ELEPHANTS

Ageing in humans- reproductive conflict
Lahdenperä, M., Gillespie, D.O.S., Lummaa, V. and Russell, A.F. (2012). Severe Intergenerational Reproductive Conflict and the Evolution of Menopause. *Ecology Letters* 15, 1283–90.

Death of granny
https://news.nationalgeographic.com/2017/01/granny-j2-killer-whale-orca-presumed-dead-washington-puget-sound/

Elephants and interest in ivory
Mccomb, K., Baker, L. and Moss, C. (2006). African Elephants Show High Levels of Interest in the Skulls and Ivory of Their Own Species. *Biology Letters* 2, 26–28.

Death of Eleanor
Douglas-Hamilton, I., Bhalla, S., Wittemyer, G. and Vollrath, F. (2006). Behavioural Reactions of Elephants Towards a Dying and Deceased Matriarch. *Applied Animal Behaviour Science* 100, 87–102.

CHAPTER 10

Surus and Hannibal's elephants
https://www.nytimes.com/1984/09/18/science/the-mystery-of-hannibal-s-elephants.html
Scullard, H.H. (1974). *The Elephant in the Greek and Roman World.* Thames & Hudson: London, UK.

Ivory in Hong Kong
https://www.admcf.org/2017/06/27/hong-kong-the-public-overwhelmingly-support-banning-hong-kongs-ivory-trade/
https://environment.yale.edu/tri/fellow/1898/

Belen Woo addressing the Legislative Council
https://www.legco.gov.hk/yr16-17/english/bc/bc06/papers/bc0620170906cb1-1393-11-e.pdf

Ivory spikes
The illegal wildlife trade in global perspective
R. Duffy – Handbook of Transnational Environmental Crime, 2016

REFERENCES

Myanmar elephants killed
Sampson, C., McEvoy, J, Oo, Z.M., Chit, A.M., Chan, A.N., et al. (2018) New
    Elephant Crisis in Asia — Early Warning Signs from Myanmar. *PLOS
    ONE* 13, e0194113. https://doi.org/10.1371/journal.pone.0194113

*A report by the International Union Conservation of Nature and Natural Resources
in 2016, based on 275 estimates from across the African continent, put Africa's
total elephant population at around 415,000, representing a decline of around
111,000 in ten years.*
https://portals.iucn.org/library/sites/library/files/documents/SSC-OP-060_A.pdf

*The great elephant census covering 15 countries, published the same year indicated
a decline of 144,000 elephants over a 7-year period, a rate of 8% a year. The
survey recorded the greatest drops in numbers in Angola, Mozambique and
Tanzania.*
Chase, M.J., Schlossberg, S., Griffin, C.R., Bouché, P.J.C., Djene, S.W., Elkan,
    P.W., Ferreira, S., Grossman, F., Kohi, E.M., Landen, K., Omondi, P.,
    Peltier, A., Selier, S.a.J. and Sutcliffe, R. (2016). Continent-wide Survey
    Reveals Massive Decline in African Savannah Elephants. *PeerJ* 4, e2354.

Zakouma National Park
https://www.smithsonianmag.com/science-nature/race-stop-africas-ele-
    phant-poachers-180951853/

DNA from tusks
Wasser, S.K., Joseph Clark, W., Drori, O., Stephen Kisamo, E., Mailand, C.,
    Mutayoba, B. and Stephens, M. (2008). Combating the Illegal Trade in
    African Elephant Ivory with DNA Forensics. *Conservation Biology* 22,
    1065–71.
Wasser, S.K., Joseph Clark, W., Drori, O., Stephen Kisamo, E., Mailand, C.,
    Mutayoba, B. and Stephens, M. (2008). Combating the Illegal Trade in
    African Elephant Ivory with DNA Forensics. *Conservation Biology* 22,
    1065–71.

Dogs picking up whale faeces
https://www.smithsonianmag.com/science-nature/meet-dogs-sniffing-out-
    whale-poop-science-180958050/

ELEPHANTS

Tusk size changes
Chiyo, P. I., Obanda, V. and Korir, D. K. (2015). Illegal Tusk Harvest and the Decline of Tusk Size in the African Elephant. *Ecology and Evolution*, 5(22), 5216–29. doi:10.1002/ece3.1769

Measuring tusk size from photos
Black, C.E., Mumby, H.S. and Henley, M.D. (2019). Mining Morphometrics and Age from Past Survey Photographs. *Frontiers in Zoology* 16, 14.

Changes in tusk size over time
https://www.nationalgeographic.co.uk/animals/2018/11/elephants-are-evolving-lose-their-tusks-under-poaching-pressure
https://www.nationalgeographic.com/animals/2018/11/wildlife-watch-news-tuskless-elephants-behavior-change/

Orphaned elephants
Goldenberg, S.Z. and Wittemyer, G. (2017). Orphaned Female Elephant Social Bonds Reflect Lack of Access to Mature Adults. *Scientific Reports* 7, 14408.
Goldenberg, S.Z. and Wittemyer, G. (2018). Orphaning and Natal Group Dispersal are Associated with Social Costs in Female Elephants. *Animal Behaviour* 143, 1–8.

Increase in poaching in Botswana
https://www.nationalgeographic.com/animals/2018/09/wild-life-watch-news-botswana-elephants-poaching/

Lifting the trophy hunting ban
https://www.nationalgeographic.com/animals/2019/05/botswana-lifts-ban-on-elephant-hunting/

CHAPTER 11

Report by the WWF that stemmed the 'Last Generation' headlines
https://wwf.panda.org/knowledge_hub/all_publications/living_planet_report_2018/

Index

287

INDEX

INDEX

INDEX

INDEX